AF758342

SHAPE MEMORY ALLOYS - PROCESSING, CHARACTERIZATION AND APPLICATIONS

SHAPE MEMORY ALLOYS - PROCESSING, CHARACTERIZATION AND APPLICATIONS

Editor

George Gibbs

scitus academics

Shape Memory Alloys - Processing, Characterization and Applications
Edited by **George Gibbs**

ISBN: 978-1-68117-644-4
Library of Congress Control Number: 2016946505

© 2017 by
SCITUS Academics LLC,
616, Corporate Way, Suite 2, 4766,
Valley Cottage, NY 10989

www.scitusacademics.com

This book contains information obtained from highly regarded resources. Copyright for individual articles remains with the authors as indicated. All chapters are distributed under the terms of the Creative Commons Attribution License, which permits unrestricted use, distribution, and reproduction in any medium, provided the original author and source are credited.

Notice

Reasonable efforts have been made to publish reliable data and views articulated in the chapters are those of the individual contributors, and not necessarily those of the editors or publishers. Editors or publishers are not responsible for the accuracy of the information in the published chapters or consequences of their use. The publisher believes no responsibility for any damage or grievance to the persons or property arising out of the use of any materials, instructions, methods or thoughts in the book. The editors and the publisher have attempted to trace the copyright holders of all material reproduced in this publication and apologize to copyright holders if permission has not been obtained. If any copyright holder has not been acknowledged, please write to us so we may rectify.

Printed in United States of America on Acid Free Paper

PREFACE

Shape memory alloys have become in the past decades a well-established research subject. However, the complex relations between properties and structure have created a continuously growing interest for a deeper insight all this time. Shape memory alloys (SMAs) are metals that "remember" their original shapes. SMAs are useful for such things as actuators which are materials that "change shape, stiffness, position, natural frequency, and other mechanical characteristics in response to temperature or electromagnetic fields". The potential uses for SMAs especially as actuators have broadened the spectrum of many scientific fields. The study of the history and development of SMAs can provide an insight into a material involved in cutting-edge technology. The diverse applications for these metals have made them increasingly important and visible to the world. Nickel-titanium alloys have been found to be the most useful of all SMAs. Other shape memory alloys include copper-aluminum-nickel, copper-zinc-aluminum, and iron-manganese-silicon alloys. Although medical applications for shape memory alloys (SMA) now dominate in today's market, there are many applications in the industrial sector which have reached large volume production that far surpass the material usage in the medical fields. In the early growth of shape memory alloy technology the most important applications were for fasteners and couplings, mainly in the military sector. With the maturing of the technology, and the broader availability of alloys, industrial applications appear in a wide spectrum of commerce. Eyeglass frames were an early example of a new use of superelasticity which has grown to be a world-wide product. Cellular phone antennas consume millions of feet of superelastic wire, and the development of underwire for women's brassieres, formerly limited to Asian market, is now expanding into a worldwide fashion.

Shape Memory Alloys - Processing, Characterization and Applications focuses on a wide range of topics from processing to the relationship between the structure and properties, as well as new applications.

The many uses and applications of shape memory alloys ensure a bright future for these metals. Research is currently carried out at many robotics departments and materials science departments. With the innovative ideas for applications of SMAs and the number of products on the market using SMAs continually growing, advances in the field of shape memory alloys for use in many different fields of study seem very promising.

TABLE OF CONTENTS

1

THE METHODS OF PREPARATION OF TI-NI-X ALLOYS AND THEIR FORMING

Radim Kocich, Ivo Szurman and Miroslav Kursa

[1] VŠB Technical University of Ostrava, Czech Republic

1. INTRODUCTION

The continuous development of science and technology in all industrial sectors means connecting and usage of a wide range of new knowledge together with implementation of new modern technologies for production of materials with high functional, specific and special properties. Intermetallic compounds TiNi with shape-memory effect are an interesting group of materials. These materials are used in a wide range of industry, such as electronics, robotics, tele-communication and also in medicine and optics. Shape-memory alloys (SMA) are a group of materials characterized by shape-memory effect (SME) and superelasticity (SE), also called pseudoelasticity.

Ti-Ni binary alloys (sometimes called "Nitinol") are probably the best known from this group of materials. Nevertheless, these alloys are not always the most suitable for the particular purpose. This factor is also the reason for seeking optimized variants of these original binary systems. One of the possible solutions is to modify binary alloys by incorporation of one or more chemical elements into the production process. The resulting materials can be summed up in the term Ti-Ni-(X), where X means presence of another element. Although the best memory characteristics are usually achieved for alloys with Ni content of 49.3 ÷ 51 at. % (Raz & Sadrnezhaad, 2004), by decreasing the content of one element (Ti or Ni) to the prejudice of the third element, modified materials are obtained, while preserving some of original characteristics. Among the main characteristics, surpassing SME and SE, mechanical properties, corrosion resistance and related biocompatibility should be mentioned (Van Humbeeck, 2001) or (Duerig et al., 1999). Intermetallic equiatomic compound of nickel and titanium thus remains as the base of modified binary materials. Nevertheless, it should be stated that characteristics of Ti-Ni SMA may be significantly modified otherwise than by the appropriate choice of chemical composition, namely by forming and thermal (thermomechanical) processing. As will be indicated later, final properties and products made of SMA are significantly influenced not only by the chosen forming technique, but also their mutual sequence. These factors together with the used technique play a major role in the manufacture of products from SMA.

2. METHOD OF PREPARATION

Production of Ni-Ti alloys is mostly done by vacuum melting, whilst various melting procedures are used (electron beam melting, arc melting (Ma & Wu, 2000) and (Meng, 2001), high frequency vacuum melting in a graphite crucible (Noh, 2001) or (Tsai et al., 1994), plasma melting, etc.). When Ni-Ti alloys are melted, there can be unfavourable effects, especially of gases such as nitrogen or oxygen. Other problems consist in the conditions suitable for crystallization and minimalization of micro- and macro-segregation connected with that. Also, contamination of the material by non-metallic intrusions has to be prevented (Schetky & Wu, 2005). Due to the formation of titanium carbide and titanium oxide in Ni-Ti, concentration of individual elements changes and thus so does the transformation temperature. Among other problems arising from the melting of Ni-Ti, there is the formation of low-melting point phase $NiTi_2$, which causes a strong tendency towards hot crack formation.

The basic requirement to metallurgy of these alloys is strict adherence to the chemical composition of the alloy, which is the main condition for obtaining the alloy with the required transformation behaviour. Another condition is obtaining an excellent microstructural homogeneity of the alloy, which is also a condition for functional reliability and guaranteed transformation behaviour. A deviation of about 0.1 at. % from the required chemical composition usually changes the transformation temperature by as much as 10 K. In Fig. 1a you can see the dependence of temperature of martensitic transformation on the nickel content in the alloy. There is a possibility of attenuation of concentration dependence of the martensitic transformation temperature by alloying with other elements, especially Cu, Fe, etc.

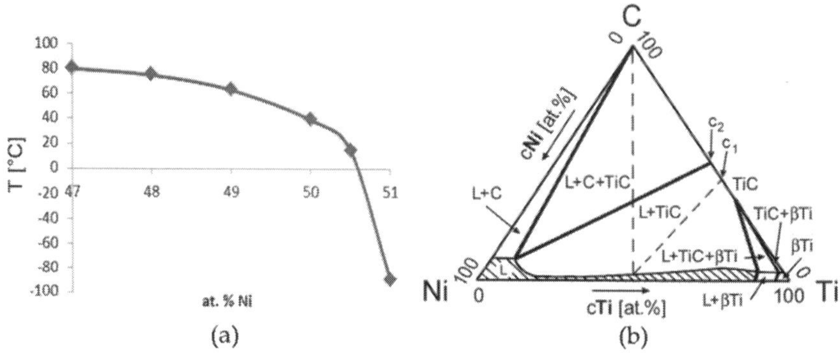

Figure 1. The dependence of temperature of martensitic transformation (a) Ternary system Ti-Ni-C (b)

Based on specific requirements of applications such as actuators/sensors, temperature control, fatigue properties, etc., various alloys with the addition of a third element giving a ternary alloy were developed (Otsuka & Wayman, 1998) or (Zhang et al. 2006).

There is a certain influence of each alloying element on transformation characteristics of the alloy. For example, the addition of Hf, Zr, Au, Pd and Pt causes the increase of phase transformation temperatures, while elements such as Fe, Co, and V have the opposite effect. Similarly, hysteresis is increased, e.g., by Fe and Nb, and, on the contrary, decreased by Cu (Ramajan et al., 2005). As a consequence of alloying by other elements, the transformation sequence is also changed; e.g., at the content of Cu below 7.7 % one-stage phase transformation B2→B19′ occurs (similarly as in a binary alloy). If the content of Cu exceeds 7.7%, two-steps transformation B2→B19→B19′ takes phase (Tang et al., 2000). The alloy properties may also be significantly influenced by alloy impurities from the production process, forming, heat treatment, etc. As it was already stated, there could be an important role of gases (O_2, N_2, H_2) and carbon. In the resulting structure intrusions of the type $Ti_4Ni_2O_x$, TiO_2etc. connected with the decrease of Ti content in the matrix can be observed. There is significant influence of these composition changes on transformation characteristics of the alloy.

Typical superelastic nitinols contain ca. 350–500 ppm of oxygen and 100÷500 ppm of carbon. The metallurgical purity (grain structure, presence of impurities etc.), of course, greatly depends on the preparation process. Ni-Ti alloys can be called high-purity alloys if they contain <100 ppm of oxygen and <20 ppm of carbon. These alloys are prepared in vacuum induction furnaces in graphite crucibles with the subsequent repeated re-melting in vacuum arc furnaces (Graham et al., 2004).

2.1. Vim – Vacuum Induction Melting

As has already been stated, VIM is one of the production processes used for the preparation of TiNi alloys. The technology of vacuum induction melting in graphite crucibles represents the existing key preparation method. Chemical homogeneity within this technology can be achieved by appropriate power control (and stirring of liquid alloy connected with that). When using this technology, the quality of the prepared alloy will strongly depend on the material of the crucible. Usually the mentioned graphite crucible is recommended – where the oxygen content can be neglected; nevertheless, carbon absorption must be considered here (there is a significant influence of carbon on microstructural characteristics and transformation behaviour). During the preparation of the material in a graphite crucible it was also found (Frenzel et al., 2004) that in the case of using Ni-pellets and Ti bars/disks the appropriate arrangement of the material in the crucible was important. The authors of this study have shown that although the inner surface of the crucible was covered with Ti disks, the content of carbon in the produced alloy was lower in comparison with the case of random arrangement of the charge. This phenomenon is caused by formation of a TiC layer, which acts as a diffusion barrier. It was also found that the carbon content strongly depends on temperature and time of dwell of the melt in the crucible. For this reason, a more intensive investigation of these effects was carried out (Zhang et al., 2006). It was established that with increasing time of dwell of the melt in the crucible the melt gets enriched in carbon.

In Fig. 1b (Du & Schuster, 1998) it is possible to see more detailed information on the isothermal section (at 1500°C-temperature recommended for melting of Ni-Ti based alloys) of the Ni-Ti-C ternary system. The composition in this system is given in atomic %. It is shown that there exists a single-phase region of liquids, extends from the area of pure Ni to the area of pure Ti. There exists only a narrow two-phase area L+-Ti which separates the area of melted material from the -Ti phase. The diagram also shows that the melted material dissolves a certain amount of carbon (this dissolution is limited). Elementary melted Ti and C cannot coexist in equilibrium state, due to this reason a TiC carbidic phase is created. The diagram in Figure 1b also predicts the existence of three phases in thermodynamic equilibrium: pure carbon, TiC carbidic phase and melted Ni-Ti depleted by Ti. The reactions between the melt will result in a melted material with higher carbon content and certain amount of TiC. In practice we cannot expect this equilibrium.

When the molten Ni-Ti enters into contact with the graphite of the crucible, inter-diffusion causes a growth of the TiC layer and the contents of carbon in the melted alloy grows. This process includes the diffusion of carbon through a thin layer of TiC which grows on the boundary between TiC/melted Ni-Ti. On the boundary between graphite / TiC and the boundary of TiC / melted material we expect local thermodynamic equilibriums. If using a pure (unused, new) crucible for preparing the alloy, the first prepared ingot will have a higher content of carbon than the next one. This fact is in accordance with the creation of the above-listed TiC diffusion barrier. It is also recommended to perform rinse-melting before melting alloys in an unused crucible.

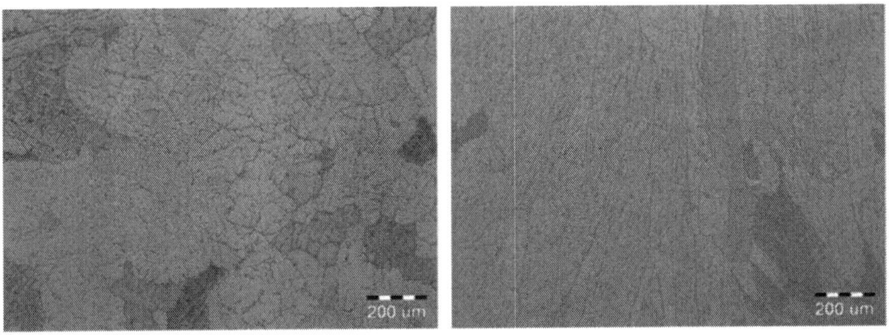

(a) (b)

Figure 2. As cast state of alloy: Ni50.6-Ti(at.%) (a), Ni46-Ti50-Co4 (b)

In order to define the exact influence of the used technique, an experimental study (Szurman & Kursa, 2010) with the aim of monitoring the influence of the preparation process on microstructural characteristics of Ni-Ti-(X) alloys was performed. The examples of microstructures of Ni50.6-Ti (at. %) and Ni46-Ti50-Co4 cast alloys are presented in Figs. 2a and 2b. As a consequence of the preparation of alloys in a graphite crucible, TiC type carbide phases are visible in the alloys' microstructure. A TEM image of the TiC phase (Fig. 3a) with the appropriate diffraction is presented in Fig. 3b. Similarly as with carbides, oxide phases can also be seen in microstructures of Ni-Ti alloys. A specific example is presented in Figs. 4a, b where particles of Ti_4Ni_2O can be seen.

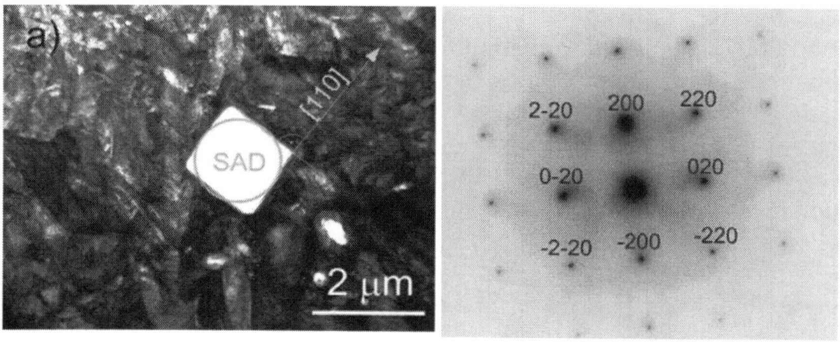

(a) (b)

Figure 3. TiC phase: TEM image (a), corresponding diffraction pattern (b)

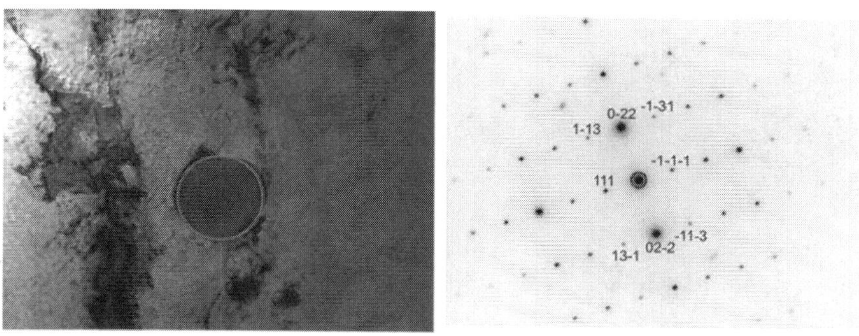

(a) (b)

Figure 4. Particle of Ti_4Ni_2O (a), corresponding diffraction pattern (b)

2.2. Plasma Melting – Plasma Furnace With Horizontal Crystallizer

This is another possible preparation process; there are, however, serious drawbacks. During this process, input elemental metals are placed in the copper water-cooled crystallizer. The crystallizer is carried by the screw below the plasma burner. Argon is used as a plasma-forming gas. For the melting as such it is necessary to use the cleanest available argon due to high affinity of titanium to oxygen. The plasma temperature during this process reaches 6500 K (Dembovský, 1985) and (Pacholek et al., 2003). The advantage of this process can be seen in the prevention of contamination of melted material by graphite from used electrodes (crucibles); high concentration of energy, high plasma flow velocity and very quick heat transfer on the heated material ensure high speed of melting. Disadvantages of plasma furnaces in comparison with vacuum induction furnaces include lower degassing of the melted metal, which depends on purity of the used argon. The key disadvantage of this process consists in insufficient homogeneity of the prepared alloy.

The development of plasma furnaces takes place in two main directions. Melting units working on the similar principle as common arc furnaces can be added to the first type of plasma furnaces. There is only one difference – that instead of electrodes, plasma burners are used and the furnace used to be equipped with special soil electrode carrying the current into the charge. The working space of furnaces is often designed to be vacuum-tight, which enables maintaining an ideal inert atmosphere. This type of furnace can be equipped with a relatively simple device for electromagnetic stirring of liquid metal.

The second furnace type is plasma furnaces with water-cooled metal crystallizers. As to the arrangement, the concept of these furnaces is similar to electronic furnaces, with the difference that instead of electron guns plasma burners are used and the furnaces mostly work with the pressure of an inert gas varying around 10^5 Pa. Exceptionally, there are furnaces with overpressure. In metallurgy, so-called low-temperature plasma in particular is considered, which is a system comprising a mixture of neutral particles with the prevailing number of electrons and positive ions with temperatures in orders of 10^3 to 10^4 K. The temperature of 10^5 K can be considered as the temperature of totally ionized plasma (Dembovský, 1978).

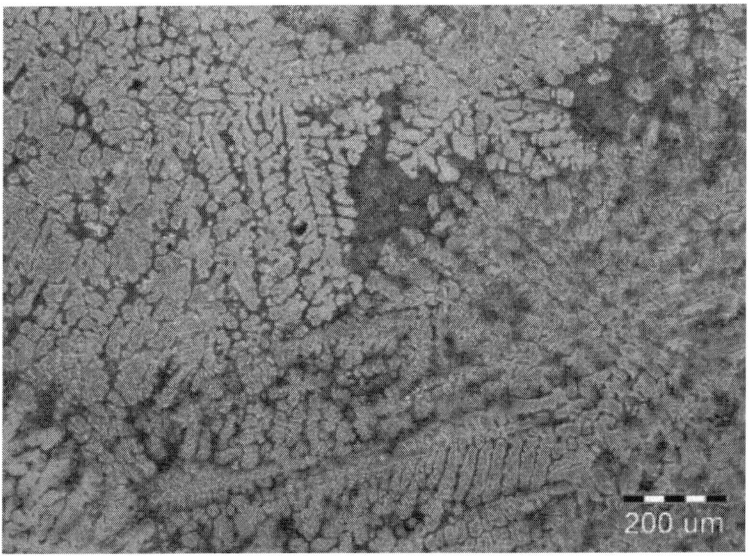

Figure 5. Microstructure of alloy Ni49.5-Ti25.5-Zr10-Nb15 (at. %), plasma

Specific experiments with melting of selected alloys Ti-Ni-(X) are described, e.g., in studies (Szurman & Kursa, 2009). Using this technique, ingots with the weight of 200–1000 g were prepared. In Fig. 5you can see microstructure of alloy after plasma melting. As you can see, the microstructure of the alloy is highly inhomogeneous. This problem is caused by very high temperature gradients during melting. At the top of the ingot the alloy is heated to a high temperature. On the other hand, the part of the ingot which is in contact with the crystallizer is intensively cooled.

2.3. Var – Vacuum Arc Melting

VAR technology is widely used to increase metallurgical purity of alloys prepared using standard procedures, e.g., in vacuum induction furnaces. This procedure is also known as "consumable electrodes". Direct current is used for the formation of arc between the electrode (melted material) and a water-cooled copper crucible/crystallizer. As a consequence, the electrode tip is melted and a new ingot is formed within the water-cooled crucible. The melt during the arc vacuum melting therefore is not in contact with the graphite crucible (as it is in the case of the VIM technique with a graphite crucible/mould), thus a "more pure" product can be obtained using this method. The carbon content usually does not exceed 200 ppm (Dautovich & Purdy, 1965) or (Wu, 2001). For this technique a very high vacuum is required.

Nevertheless, there are also drawbacks of this technique – small volume of the alloy and low convection in the melt which may cause inhomogeneity of ingots. That is also the reason why this procedure is usually repeated several times. Often the VIM+VAR process is applied for the preparation of Ti-Ni-(X) alloys. VAR technology is also preferred for preparation of experimental material for basic research of Ti-Ni-(X) alloys. For example, the study (Choi et al., 2005) describes the preparation of experimental Ni-Ti alloys alloyed by Fe. In this study, the ingot prepared using this method is homogenized at 1273 K for 24 hours. In another study (Sakuma et al., 2003) heat treatment at 1223 K for 1 hour after the preparation of the material using this method is proposed. The specific regime is also mentioned in the study (Jung et al., 2003) where heating at 1100°C for 100 h was used.

2.4. Ebm – Electron Beam Melting (Vertical Zone Melting)

Crucible-free zone melting – or the floating zone process known as the FZ method represents another specific preparation method. The formation of a narrow melted zone is performed using electron heating. The melting takes place in a vacuum, and values of 10^{-2} Pa are reached in this technology compared to 10 Pa in VIM technology. Using the method of electron zone melting with suitable oriented nuclei, even monocrystals of many high-melting metals – W, Mo, Ta, Nb, V, Zr, Ti, Re – can be prepared. The zone is maintained in the floating condition mainly by forces of surface tension. The zone stability depends on gravitation, surface tension and density of the melt, on material composition and also on direction of zone movement. To maintain the stability of the zone, an outer magnetic field with so-called supporting frequency is used (Kuchař & Drápala, 2000). A circular shape ingot prepared in a vacuum induction furnace is used in this case as input material. There is no risk of other contamination of the material with carbon in this technology (there is no crucible). Carbon contents are usually lower than in the case of the alloy preparation using VIM technology. Contents of gases are usually low as well because of using a high vacuum. The contamination with gases thus depends only on the quality of an input casting and tightness of the vacuum system. The disadvantage of this technology is the control of chemical composition – evaporation of some elements can be expected here during melting. Another drawback is the rather small volume of the prepared material; therefore this technology is not suitable for commercial use (Ramaiah et al., 2005).

Rather integrated results regarding (un)suitability of various methods of preparation are presented, e.g., in a recent study (Szurman & Kursa, 2009). The aim was an intercomparison of techniques selected for the preparation of Ti-Ni-(X) alloys from the point of view of the microstructure and gas contents in the material. Also in this situation a distinct decrease of gas contents after the preparation was observed here. The specific microstructure of the prepared alloy Ni50,6-Ti (at.%) is presented in Fig. 6.

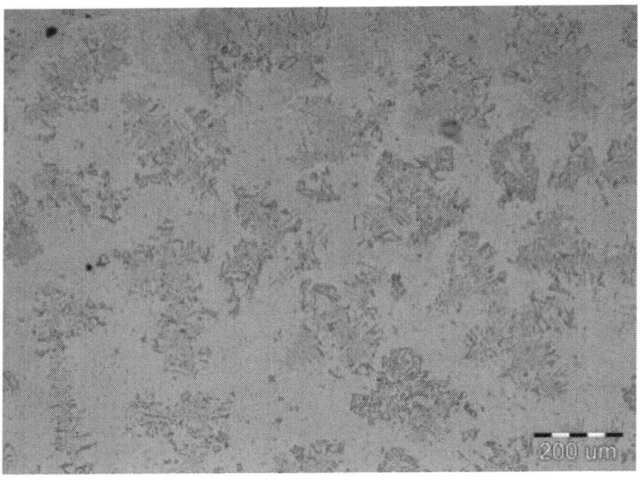

Figure 6. Microstructure of alloy Ni50.6-Ti (at. %), EBM

2.5. Preparation Of Alloys Using Powder Metallurgy

Powder metallurgy is an important and suitable method for the production of the mentioned alloys. Methods of atomization were developed for preparation of powder metals with precise control of composition. However, the biggest problem with these alloys is oxygen and carbon content. The content of oxygen can be up to 3000 ppm, but it can be decreased by careful treatment to 1500 ppm (Schetky & Wu, 2005). It is well known that with increasing content of impurities (especially oxygen and carbon), transformation temperatures are decreased and a brittle secondary phase is formed (Mentz et al., 2008). Subsequently, the composition of NiTi matrix (depletion by Ti) is significantly influenced by oxide and carbide intrusions and thus can cause degradation of functional and mechanical properties, which was also confirmed, e.g., in the study (Mentz et al., 2006). Other methods are hydridation, pulverization and mechanical alloying (Wu, 2002).

The method for preparation of Ni-Ti alloy using powder metallurgy is described in the study (Mentz et al., 2008). At first, the alloy was prepared by authors using a classical VIM melting from high-purity input raw materials, then atomization with 6N argon followed. The obtained powders were sealed into evacuated capsules made from stainless steel and then compacted using the HIP method. It was also found that each technological step is accompanied by increased content of impurities (oxygen and nitrogen). In the study (Bertheville & Bidaux, 2005) the authors performed the preparation of Ti-rich Ni-Ti alloy from elementary powders – Ni and TiH_2. In another study (Zhu et al., 2005), preparation of Ni-Ti alloy by sintering in argon from elementary Ni and Ti powders is described. Another method of preparation is then described in the study (Mousavi et al., 2008), where a method of mechanical alloying is described. The Ni-Ti alloy was prepared from elementary powders in a planetary ball mill under atmosphere of Ar. During sintering of Ni and Ti, a significantly exothermic reaction takes place so that the heat generated during the process is used for the formation of the intermetallic compound TiNi.

Elemental powders can also be sintered using "combustion" synthesis or explosion. In the first case, a laser can be used as an external energy source (Bertolino et al., 2003). As for explosive sintering, the reaction takes place after explosion during the temperature rise. Another method is based on the passage of an electric current of suitable value under optimal voltage. It was found that the optimal current density is 2822–5290 $kA.m^{-2}$. The observed sintering times were within the range 5÷40 min (Locci et al., 2003). Ni-Ti materials prepared from powder metals are very porous and contain other intermetallic compounds such as Ti_2Ni and Ni_3Ti.

3. FORMING OF SMA

Apart from the already mentioned influence of the selected preparation method of SMA, final properties and behaviour of SMA will be also determined by next processing including heat treatment. This means not only the chosen method of forming, but also a sequence of given forming operations. Also the influence of applied regimes of heat treatment should be considered. All the mentioned factors have their own partial effect in formation of final properties of SMA-based materials. With admixture elements, such as Cu contained in binary NiTi alloys, martensitic transformations are considerably shifted and at the same time also mechanical or thermomechanical characteristics of SMA are changed, as can be seen, e.g., in the study (Liu, 2003). Hand-in-hand with forming or heat processing, strengthening and healing processes are also important. Several studies are known, e.g. (Gili et al., 2004) or (Morgiel et al., 2002), which confirm the importance of chemical composition (Cu content) during dynamic recrystallization, esp. near borders of grains. On the other hand, other studies (Nam et al., 1990) bring information on formability of SMA when Ni was substituted by Cu. If Ni is substituted by Cu up to the content of ca. 10% (i.e. binary NiTi alloy will be modified to Ti-40Ni-10Cu), then the phase transformation will take place in two steps and these alloys will be much more deformable in the martensitic state than original NiTi alloys. Despite these findings it is still valid that when Cu content exceeds the limit 10 at.%, Ni-Ti-Cu alloys exhibit a rather low formability.

The chosen method of forming together with the method of heat processing is directly proportional to achieved characteristics. Although SMA are mostly used in the form of thin belts, wires or pipes (Kursa et al., 2005), all these products are produced by forming from original cast ingots. Ingots in the cast state are characterized by a very low formability and usually only a small or no memory or superelastic effect. With subsequent hot or cold forming these properties are modified. In Fig. 7 you can see a scheme of basic distribution of forming operations which are usually used for plastic deformation of SMA. Similarly as in other materials, the main aim of hot forming is to change the dimensions and shape of cast ingot, together with modification of its unfavourable microstructure. For the "destruction" of the original dendritic structure, some deformation depending on the cross section of the treated ingot should be applied. It is not unusual for the real size of the applied strain which will provide the required changes to reach values of around 90%. High degrees of deformation performed as hot forming are often also conditioned by requirements arising from the consecutive cold forming, during which such significant reductions of cross section cannot be realized (Ramaiah et al., 2005). Nevertheless, especially recently, when an explosion of unconventional forming techniques occurred, it can be said that intensities of applied strain can reach, and in practice do reach, much higher values than 100%. It should be stated that in these cases it is a shearing deformation where no significant changes in cross section occur. The main aim of these unconventional forming techniques is to achieve structural modification with the effort to deform materials at temperatures as low as possible.

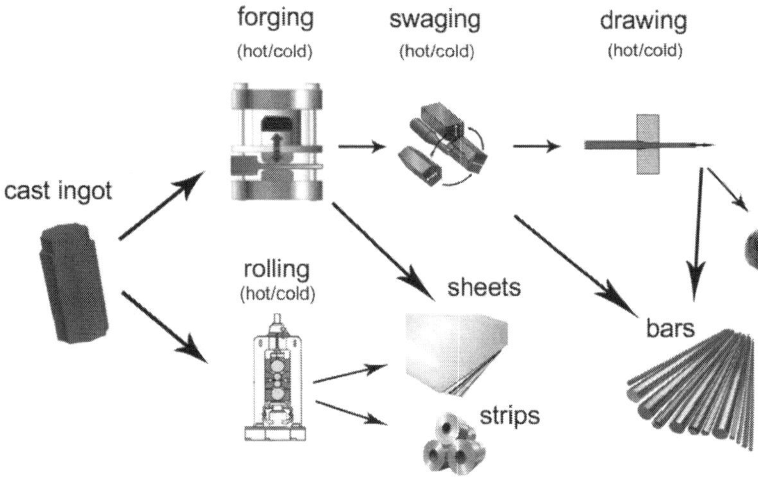

Figure 7. Scheme of basic forming operations used for plastic deformation of SMA

The conventional treatment (forming) of SMA is usually divided into more stages. A frequent sequence of individual operations consists in melting, casting, hot swaging, cold rolling and drawing. Especially during cold forming techniques it is common to insert heat treatment between partial operations. So it is obvious that the transformation behaviour of a particular alloy will be influenced by each of these mentioned operations. In the first stages (melting, casting), there is an already mentioned factor of chemical composition. However, the production process itself can be performed in several various ways with different influence on the studied characteristics, which is documented by a high number of performed studies (Frenzel et al. 2004 and Zhang et al., 2005 and Frenzel et al., 2007a). Regardless of the chosen technique of melting, increased attention should be paid in all techniques to minimization of additional elements, especially oxygen and carbon. These elements have negative influence on the memory effect and also on the brittleness of the particular alloy, which is not without perceptible consequences, especially during the stage of forming.

During alloy forming, which is usually performed in the temperature range 300÷900°C, in addition to the present admixtures, defects in the crystal lattice also begin to come to light. To be more specific, both point defects and changes in the dislocation density begin to activate, which will significantly influence healing and precipitation processes (Frenzel et al., 2007b) or (Kocich et al., 2007). In binary NiTi alloys at around temperatures of 400°C the softening process begins, while at temperatures of 900°C formability (elongation) of alloys determined by tensile tests can exceed 100%. Although SMA at these temperatures exhibit relatively good formability, during alloy forming some cracks may appear, especially near the edges. Usually bars or plate slabs are prepared by forming (forging, rolling). The deformation behaviour of SMA can probably be considered optimal in the temperature range near 800°C. Just these temperatures lie in the range where the alloys are workable and at the same time oxidation of their surface is not as massive as at higher temperatures (Wu, 2001). If forming temperatures are too high, the mentioned oxidation takes place and with increasing temperature the degradation of the material increases as well. The consequence of these processes is formation of very stable oxide layers which are often a part of the surface and cause destruction of the material due to the formation of cracks. When SMA is heated to the temperature 900°C, the alloy tends to be brittle because of occurrence of the Ti_4Ni_2O phase.

After hot forming of alloys, heat treatment is very often applied. The influence of annealing temperatures or cooling rate after annealing is already known very well; more detailed information on partial modes can be found, e.g., in the study (Standring et al., 1980). Generally it can be said that longer annealing times cause higher A_f temperatures (austenite finish). It is known that an increase of transformation temperatures (A_s, A_f) depends on the technology used for the preparation (machining) of SMA and is caused by temperature-induced stresses and defects. High M_s temperature is attributed to incomplete transformation during heating. The mentioned heat treatment is used for optimization of physical and mechanical properties together with maximization of shape memory effect and pseudo-plastic behaviour. The main reason for performed heat treatment after forming is thus modification of transformation temperatures for specific applications. When compared to the relatively wide interval of forming temperatures, the range for annealing temperatures is considerably narrower (300° – 525°C). The times used for annealing are usually in the order of minutes (5–30). Relatively stable resulting transformation temperatures are documented for annealing at temperatures of about 500°C and times shorter than 10 minutes (Liu et al., 2008). With increasing time of annealing A_f temperature increases and stress decreases. Increase of A_f is usually accompanied by depletion of the NiTi matrix by Ni, which is precipitated in the form of precipitates.

3.1. Swaging

As has already been mentioned before, one of the first forming procedures used for plastic deformation of cast ingot is swaging. Swaging is a forming process characterized by a very high rate of deformation (i.e. potential possibilities for the production are 4–6 pieces per minute). It is usually performed on swaging machines. There are many significant advantages of swaging, among them, e.g., possible high reduction of cross section at relatively low energetic costs, significantly dimensionally more precise forged pieces, higher surface quality and considerable improvement of mechanical properties of these products. This process has been known about for a relatively long time, which is also documented by many published studies. One of the main aims of these studies is to find a method of detection of pressure distribution on contact surface during individual deformations (Zhang, 1984). Individual approaches consider various assumptions related to the parameter of applied energy (Choi et al., 1997and Canta et al., 1998) or to the course of metal flow (Standring et al., 1980) or (Wang et al., 2005).

Swaging can be divided into two main groups. The first group is hot swaging; the second one cold swaging. Both named groups are procedures characterized by high efficiency and they can be used both for the production of full bars and differently shaped pipes. Swaging is a process which can be described using gradual (incremental) deformations, and is widely applicable for the production of engineering parts such as disks, rings, gear shafts, etc. The main difference between cold and hot swaging is in the working temperature of the process and also the used lubrication (in cold swaging).

However, it is generally known that high-temperature swaging, when compared to the original state of SMA after homogenization, need not necessarily cause a significant difference in reached transformation temperatures. On the other hand, e.g., after cold rolling, there is a distinct (usually full) suppression of phase transformation. It should be noted however that even short annealing of such a deformed state will be enough to restore the memory effect. The annealing will unblock obstacles which prevent the mentioned transformation process and the transformation can proceed again. Although the values of transformation temperatures in the state after hot forming and in the original state are not very different, there are considerable differences in the structural arrangement. During forming, a distinct decrease in grain size occurs, which also determines the final mechanical properties of individual states. Based on the size and type of applied strain, this value after forming can be increased to twice the original value.

3.2. Rolling

During hot rolling, individual grains of SMA forming the structure are deformed and simultaneously recrystallized, which preserves their equiaxed microstructure. At the same time the deformation reinforcement is compensated. The source material is usually in the form of cast semiproducts such as plate slabs, bars or ingots, or semiproducts after forming (e.g., by previous swaging, etc.). Heating of SMA is usually performed in electric furnaces, since SMA semiproducts are generally of small dimensions. In contrast to swaging, there is not such a massive generation of deformation heat to maintain the working temperature of formed components or its slight increase. Although during rolling the temperature as a consequence of deformation changes depending on the deformation rate, it is not enough to compensate the heat loss to the environment or into the tools. That is why for most SMA products (semiproducts small in dimensions) there is a serious danger of going under the recrystallization temperature during the hot rolling itself. The temperature of formed semiproduct after rolling should be near the range from 50 to 100°C above the recrystallization temperature to ensure sufficient heat for the process. In the event of failure to comply with this condition, intermediate heating must be performed before the subsequent reduction. There are relatively many published experiments in the field of rolling focused on the effect of ausforming (forming in the region of austenite) or marforming (forming in the area of martensite). For example, the deformation behaviour of binary NiTi alloys during hot pressure or hot tensile tests is mapped (Dehghani & Khamei, 2010 andMorakabati et al., 2010). Suzuki in (Suzuki et al., 1999) states that significant increase of hot formability can be achieved in NiTi alloys by forming at temperatures of 900° – 1000°C, but there is the drawback of the above mentioned surface oxidation.

When hot rolling is carried out, cold rolling usually follows. Cold rolling of SMA is a process which is much more difficult than the same hot process. The main factor that complicates this procedure is the absence of healing processes which are activated at increased temperatures. During cold rolling, in the course of deformation solidification of SMA takes place and consequently formability of materials decreases. Other influences include a higher value of deformation resistance when compared to hot rolling. Just as a consequence of the limited formability of SMA, microcracks may appear during cold rolling in rolled products. It should be stated that it can occur even at low reductions of height (~20%). In cases when the final wire is produced by cold rolling in calibres (with diameters usually lower than 5 mm) the forming should take place in more stages. That is why a higher amount of passes is necessary to obtain the final wire. Probably the main reason for usage of cold rolling is to obtain dimensional accuracy and also the high surface quality of products prepared in this way. A side effect of cold forming (rolling) is suppression of the shape memory effect, while there is an increase of strength properties and a decrease of plastic properties.

With increasing content of Ni, rolling is more and more difficult and when the limit of 51 at.% of Ni is exceeded, any rolling of NiTi alloys is extremely difficult. The main reason is considerable deformation reinforcement. As is obvious from the more experimental results, yield strength of annealed NiTi alloys is usually lower than 100 MPa, but already after deformation of 40% this limit increases to values of about 1000 MPa. If the deformation continues, it would be very probable that cracks in the material would appear or even destruction of the material would take place. Intermediate annealing must therefore be performed before the next forming. This will cause a decrease of strength properties and a partial recovery of plastic properties. The just mentioned combination of deformation-healing (annealing) causes refinement of the final structure as a result. The temperature of intermediate annealing and its length will accordingly be the key parameters for the microstructure development. Generally it can be said that the temperature of intermediate annealing is lower than the temperature necessary for hot forming and usually is about 600°C (Wu, 2002).

It is known that during cold forming (rolling) the increase of the volume fraction of martensite is much higher than during hot rolling. Relatively many experimental studies focused on thermomechanical processing of SMA and mapping the effects of rolling and subsequent annealing, e.g. (Kurita et al., 2004), confirm this knowledge. It can be briefly stated that the higher the annealing temperature is, the lower content of retained martensite you will find. However, this fraction can be present in SMA even at relatively high temperatures, which is also documented in numerous studies (Brailovski et al., 2006and Lin & Wu, 1994). The reason is the presence of dislocations which slow down its conversion to austenite during post deformation annealing. The potential softening during annealing can be considered in three ways (mechanisms): a) dislocation recovery after which retransformation of martensite to austenite takes place (at temperatures of about 400°C), b) particle stimulated nucleation (PSN) at middle temperatures (~500°C), c) recrystallization of the matrix at high temperatures (~600°C). The needed level of deformation ensuring stable shape memory and superelastic behaviour of SMA (during cold rolling) is usually above 90% (Kim et al., 2006).

Among other drawbacks of rolling, there is only a small possibility of controlling the grain morphology or texture at adequate refinement of the microstructure. These are substantial effects which can influence the shape memory characteristics. Generally it can be stated that in the case of the requirement to maximize the shape memory effect, the best solution is to use the NiTi alloy in the state after hot rolling. In the case of the requirement to obtain high strength and hardness at acceptable reversible deformation, the NiTi alloy should be used after cold rolling.

These were not the only reasons for the impulse for researching the effect of unconventional forming techniques on deformation behaviour, or transformation characteristics of SMA. Especially in recent times there is a significant effort during research regarding the application of Severe Plastic Deformations (SPD) in memory materials. Among other applied techniques within the group of SPD, there is e.g. High Pressure Torsion (HPT) technology, or the Equal Channel Angular Pressing (ECAP) process. It was already confirmed several times by experiments that these techniques are a very effective tool for influencing the transformation characteristics, cyclic stability SMA and simultaneously relatively easy control of the texture of these formed materials (Kockar et al., 2010).

3.3. Spd Processes

As has already been stated, most of the characteristics of SMA are mainly based on reverse martensitic transformation, which is controlled by chemical composition, microstructural parameters and also the method of preparation. The possibility to control functional properties of materials based on Ni-Ti alloys using thermomechanical treatment can be improved through microstructural refinement, which is documented in many studies, e.g. (Sergueeva et al., 2003). It is most desirable to obtain the structure characterized by very small grain (subgrain) size – ultra fine grain (UFG) structure. Nevertheless, it is necessary to simultaneously preserve exceptional properties of memory alloys.

It is known that TiNi-based materials characterized by a very small grain size can be prepared via three methods. The first method is chilling of cast. The second one consists in the preparation using SPD methods and the third one is then the combination of conventional techniques and subsequent annealing. Using the first and the second method enables one to obtain a fine-grained structure with the grain size in the range of 200–600 μm. If conventional forming with subsequent heat treatment will be used, then generally a larger grain size can be achieved compared to both of the previous variants. Nevertheless, a significant advantage of this method is obtaining a relatively equiaxed structure, although with a little larger grain size than in the case of SPD. A distinct difference between the conventional and unconventional method of forming is mainly in the intensity of applied strain. During conventional forming, only limited degrees of deformation can for technological reasons be applied on the formed material during one forming cycle. In addition to that – as has already been discussed above – this process should usually be performed at higher working temperature.

The principle of SPD processes is based on repeated application of high degrees of plastic deformation during individual forming cycles. This fact, together with relatively low applied temperature when these processes occur, is behind the structural refinement during SPD. The result of such deformations is equiaxed structures characterized by a relatively high amount of grains with high angle misorientation. In addition to these features, the structures can also be described by the presence of subgrains with high dislocation density (especially after ECAP). However, it must be noted that the materials prepared using SPD also contain areas with high internal stress, at least when compared to cast materials. In the case of HPT, it is possible to obtain even amorphous states which, after subsequent annealing, are transformed into nanostructural arrangements. Based on the selected regime, the final grain (subgrain) size can be expected then. The applied annealing of SMA is usually in the temperature range of 200÷400°C. This is, at the same time, the interval of relatively stable grain size, nevertheless there is a significant decrease of dislocation density inside grains. With increasing temperature (~500°C) the grain size is increased to twice the original value, while grain boundaries are also much better defined. The reason of this increase is in the course of healing processes.

3.3.1. High Pressure Torsion

Localization of applied strain can thus be achieved using large plastic deformations. One of the first suggested and so far the most effective SPD method is the high pressure torsion process (HPT) (Fig. 8). It should be stated that in spite of certain limitations, deformation behaviour of a wide range of materials can be studied with this process. The consequence of the application of SPD is the destruction of the crystal lattice with subsequent transformation to the amorphous state. Accumulated dislocations or grain boundaries are the main driving force of the amorphization process. High density of dislocations may cause formation of amorphous bands thanks to shear deformation instability. In SMA, especially in NiTi-based alloys, a strong crystallographic texture is formed (Frick et al., 2004). In the following course of the process, refinement of grains and simultaneously amorphization are observed. This continues up to the full amorphous state of the formed material volume. According to this study and others, even the nanostructure formed like that contains characteristics of the texture, while in this state no directional deformation occurred. Individual nanograins exhibit preferred orientation which corresponds with orientation of nanograins preserved in the structure after HPT. A certain explanation may be knowledge assigning this influence to nanocrystals thanks to which heterogeneous nucleation occurs.

It should be noted that such a structural state is stable only at low temperatures. For these reasons HPT is usually performed at room temperature or even lower. However, findings on the dependence between M_S temperature and formation of nano-crystalline structure in SMA materials based on NiTi are very important. If the deformation temperature during SPD is lower than the corresponding M_S temperature of the formed alloy, then there is a high probability of formation of nano-crystalline structure. If the deformation temperature is between the M_s temperature and the highest temperature of the beginning of martensitic transformation influenced by deformation (M_d), then the probability of the nanostructure formation is lowered. If the deformation temperature is higher than the M_d temperature, the probability of the nanostructure formation is very low. If after SPD post-deformation annealing is applied, then a submicrocrystalline structure is formed.

It was already mentioned that at specific temperature regimes set during annealing the alloy can crystallize into grains with the size (10–40 nm). The specific temperature range where NiTi alloys are stable is 250° – 300°C (Prokoshkin et al., 2005). With increasing temperature of annealing, the final nanostructure will be "coarser". Total amorphization of the structure is usually achievable only in cases when the deformation temperature is lower than the martensite start temperature (M_S). Rate of the structure amorphization is influenced to large extent by the applied pressure during HPT. Higher imposed pressure suppresses the tendency to form an amorphous structure from nanostructure and as well to form nanostructure from deformably reinforced dislocation structure, as confirmed by the study (Prokoshkin et al., 2005). The probable reason is a decrease of M_s temperature, due to higher values of pressure deformations.

Materials processed using HPT and subsequent heat treatments usually reach very high strength properties (strength higher than 2 GPa). Surprisingly, relatively high plastic characteristics are also preserved (elongation at break up to about 40%) (Sergueeva et al., 2003). There are also assumptions to obtain super-plastic behaviour of NiTi alloys. Although the HPT technology appears to be a suitable candidate for positive modification of the properties, it is in principle excluded from commercial use by its main drawback (very small samples)-see Fig. 8b.

3.3.2. Equal Channel Angular Pressing

It was necessary to process larger volumes of products, which has led to looking for alternative ways of HPT substitution. One possible candidate for meeting the scheduled targets appears to be the equal channel angular pressing process (ECAP). In Fig. 9a you can see the assembly for practical application of this technique. It must be noted that, in contrast to HPT, various shapes of material sections can be processed here (Fig.

9b). As it was already mentioned, there is an effort when using SPD techniques to decrease the working temperature, because it is known that with decreasing deformation temperature the final grain size in the final structure of the treated material also decreases. Since SMA are characterized by a relatively high deformation resistance, the ECAP process – in contrast to HPT – should be performed at temperatures relatively higher. The most suitable range of forming temperatures found in experiments is 400° – 500°C. Amorphization of the structure, however, cannot occur in this temperature interval. This fact also determines higher size of final grains when compared to HPT technique. The size of the grain is usually decreased to a value that lies in the region of 200–300 nm. Although even in this case the strength properties are significantly increased while preserving relatively good plastic properties, these values are lower when compared with the state after HPT. Also the combination of ECAP and thermomechanical treatment is among tested procedures for another reduction of the grain size. To be more specific, cold rolling was applied after previous angular pressing (Pushin et al., 2006).

(a)

(b)

Figure 8. Machine for HPT process (a) detail of processed sample (b)

NiTi alloys are very sensitive to the exact chemical composition, which is manifested even during their deformation using SPD techniques. For example, in the study (Khmelevskaya et al., 2001) the binary $Ni_{50}Ti_{50}$ alloy absolved 12 passes (ECAP) at the temperature 500°C, or 8 passes at the temperature 400°C; in the case of $Ni_{49.3}Ti_{50.7}$ alloy, only 3 passes could be applied, since the material exhibited relatively low formability and it was destroyed.

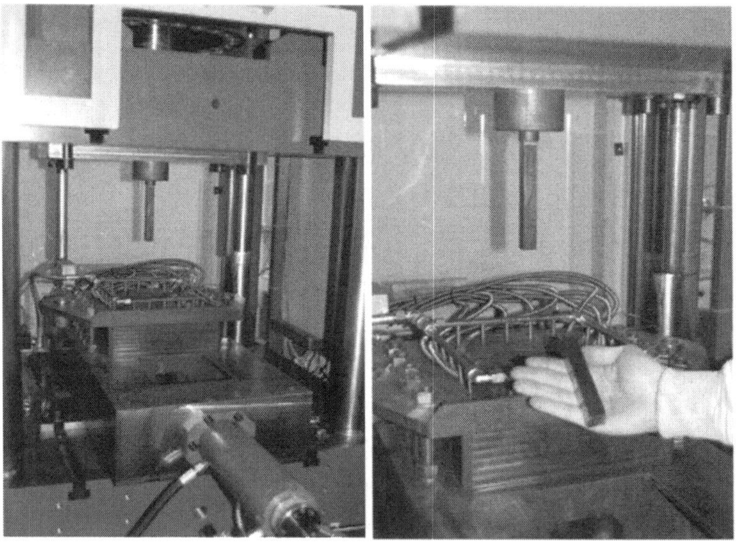

(a) (b)

Figure 9. Machine for ECAP process (a) detail of assembly and processed sample (b)

The reason for the formation of a fine-grained structure is probably the proceeding recrystallization process which causes gradual increase of disorientation of subgrain boundaries up to the formation of high-angle grain boundaries (HAGB). As to the ability of reversible deformation, its value after ECAP is comparable to the value obtained after application of conventional forming techniques (cold rolling) followed by heat treatment. It is documented, e.g., by the already mentioned study (Khmelevskaya et al., 2001) where the value of such deformation was determined to be ~7%. It should, however, be noted that the temperature interval of the memory effect after ECAP is narrower and also with lower values than after cold rolling. Similarly as after HPT, the temperature stability of SMA by annealing was tested up to temperatures of 500°C. Similarly, also after ECAP, SMA appeared to be stable, but significant reduction of dislocation density occurred. Increasing temperature caused an increase of grain size.

Generally it is known that cooling of NiTi alloys below the M_s temperature leads to higher formation of coarse-grained structure (50÷80) μm of the R phase. As confirmed by the study (Pushin & Kondratjev, 1994), cooling of coarse-grained monocrystals of B2 NiTi alloys below the M_s temperature leads to the formation of R-martensite with rhombohedral (or hexagonal) lattice. In the case of occurrence of other admixtures in binary and multiphase alloys, formation of monoclinic B19′martensite takes place during cooling below the M_s temperature. In particular, occurrence of elements such as Cu, Pd, Pt or Au during cooling of the alloy below the M_s temperature leads to formation of orthorhombic martensite B19 (Pushin, 2000). Generally it can be said that martensitic transformation on the microstructural level is caused by the presence of microtwins, while on the level of internal areas of grains it is caused by the formation of coherent crystals.

Martensitic phase transformation thus usually takes place from cubic B2 high-temperature phase (austenite) to monoclinic B 19′ phase (martensite). This process is accompanied by high deformation. According to the results from the experiment (Waitz et al., 2004), if the grain size in SMA is in the region of nanometers, then the high density of grain boundaries will act as a significant obstacle during mentioned transformations. There are two main reasons why martensitic transformation in nanomaterials is suppressed. The decrease of transformation temperatures depending on the decrease of grain size follows from the mentioned study. It was also proved that full suppression of martensitic transformation takes place when the grain size is smaller than 50 nm. From the point of view of phase stability, these small grains have a significant influence on morphology of martensite (Waltz, 2005).

Thanks to the experiment performed with the Ni50.4Ti alloy, the specific influence of forming on transformation characteristics and also on microstructure development was mapped. The experiment describes the problem of combination of conventional forming techniques together with unconventional ones. For forming, SMA characterized by the content of O_2 (0.0624 wt.%) and N_2 (0.0039 wt.%) was used. The content of carbon (0.055 at.%) was determined using spectrometry. The diameter of the cast was 20 mm and the cast length 350 mm. Then homogenization at the temperature of 850°C is followed by subsequent cooling.

The forming itself was suggested in the first phase using swaging then pressing using the ECAP technique was performed. Swaging was performed at the temperature of 850°C. During swaging, the strain was applied gradually in individual reductions. The total strain applied on the cross section was 66 %. On the contrary, the ECAP technology was applied on SMA at the temperature of 290°C. Because of possible "negative" influence on the temperature fluctuation of the formed material, the pressed samples were placed in the steel "cans". For the ECAP, matrix with the angle 105° between individual canals was used; the extrusion speed was set to 1 mm/s. The extrusion itself consisted of two performed passes, where Bc was chosen as the deformation path. In particular, the influence of intensities of applied strain in relation to mechanical and thermophysical properties and the course of healing processes were studied. To be able to determine the mentioned effect, heat treatment was carried out after the performed deformation. In addition to optical microscopy, RTG diffraction was also used for evaluation of changes. The differences between after swaging followed by ECAP are obvious from the attached photos (Fig. 10).

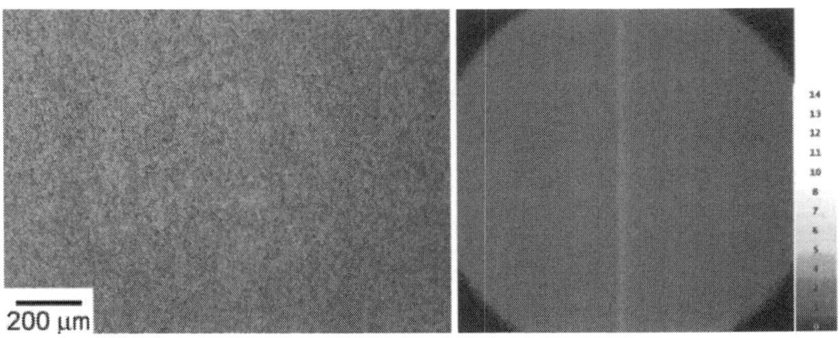

(a) (b)

Figure 10. Ni50.4Ti alloy (swaged+ECAPed): microstructure (a), diffraction fringe (b)

To be able to specify the differences caused by applied strain, it was necessary to perform heat treatment of the deformed materials. These annealing using selected temperature regimes (550°C/15 min, 600°C/15 min, 650°C/15 min, 900°C/15 min) should provide information on the influence of the accumulated deformation on the beginning of corresponding processes, especially recrystallization.

The azimuthal profile of diffraction lines of the NiTi alloy confirms that all mosaic blocks (DCA-diffraction coherent areas of crystallites) in the structure after ECAP are smaller than 10 μm (Fig. 10b). The DCA are regions that scatter coherently. These areas are defined by borders where is high dislocation density. Within the framework of the mosaic blocks, the moving dislocations face minimal resistance. Inside the mosaic blocks is much less of dislocations than on the borders of blocks, that is why the mosaic blocks are called as dislocationless cells (areas) in the crystallites. Nevertheless, this predication is not quite accurate because one of obstacle in these blocks for dislocation movement is Peierls – Nabarro stress (i.e. dislocation is moving through the mosaic block, so that mosaic block must have at least one dislocation). This labeling (dislocationless blocks) is commonly used, important factor is that inside of the mosaic blocks is considerably less of dislocation than on its boundaries. Due to this matter (low dislocation density inside of mosaic block) can X-ray be diffracted coherently (that is why the mosaic blocks are also often marked as Coherent Scattering Regions – CSR´s). The presence of blocks smaller then 10 μm was confirmed by XRD. Generally may be noticed that mosaic blocks (obtained by XRD) are not the same as grains observed by optical microscopy. Mosaic blocks can be seen by presence of reflections on XRD patterns. In the case of the large block presence will be its reflection large as well. Since cannot into the diffracted volume be much of big mosaic blocks present, there will be their reflections on diffraction lines clearly separated. If will be mosaic block small there will be much of their reflections in diffracted volume. That is why individual reflection are overlapping (they are not separated i.e. reflection is continuous). More detailed information about this technique can be found for example in (Hindelah & Hosemann, 1988).

As a consequence of recrystallization processes after annealing of the extruded alloy at temperatures of 600°C and higher in the structure blocks larger than 10 μm were formed, while after the same length of time annealing at lower temperatures all mosaic blocks remained smaller than 10 μm. It should be noted that the size distribution of mosaic blocks was bimodal (i.e. blocks larger than 10 μm and blocks smaller than 10 μm). This means that coarser crystallites contain more nickel than smaller crystals. The reason is the asymmetry of the area in the phase diagram of titanium – nickel formed by a solid solution of TiNi. Microinhomogeneity of the alloy also influences mechanical properties of the alloy. The dispersion of such elementary composition and thus lattice parameters of individual mosaic blocks in the structure causes so-called microstrain (strain of the 2nd order) influencing the dynamics of dislocations, but also nucleation and growth of microcracks. These strains of the 2nd order also influence shifts of walls of mosaic blocks and thus the course of recrystallization of the alloy. The authors of the experiment performed with the Ni50.6Ti alloy also arrived at similar results (Kocich et al., 2009).

It means that even with very thoroughly conducted preparation of material based on shape memory alloys, microinhomogeneities may be formed. As concerns the forming technologies as such, it can be stated that after completed rotary forging (swaging) the initial grain size and therefore also the ensuing properties were substantially changed. The ECAP process also proved its influence on the final appearance of the micro- and substructure. The differences are even apparent in the number of passes; if material passed through the matrix only once, then no mosaic blocks smaller than 10 μm occurred in the structure, however, influence of the second pass can be seen here, when the samples of the NiTi alloy already contained bi-modal composition. The temperature that appeared to be the starting temperature for initiation of re-crystallisation was 600°C.

4. CONCLUSIONS

For some specific purposes, the exclusivity of binary shape memory alloys should be modified by the addition of another element. These prepared Ti-Ni-X based materials are currently very progressive materials. As has already been indicated, their final properties will, to a great extent, be determined by the chosen element. To a great extent, it is thus the factor of chemical composition; nevertheless that is not the only one. A significant influence in the sense of the effect on transformation characteristics together with other properties was also demonstrated for the preparation method. There is an outstanding dependence on used forming technology and their relationship or thermomechanical conditions play an important role. During conventional forming, the deformation behaviour of SMA binary systems is relatively well known. As seems to be the case, multiphase SMA are very sensitive to even small deviations from the required chemical composition. These nuances can be observed through the shifts in recrystallization temperatures, which also means the shifts in their optimum forming temperatures.

Particularly at this time there is intensive research of use of unconventional forming techniques for SMA. Although it is already known that when using these procedures, in contrast to conventional forming, it is possible to obtain structures characterized by grain size of nanometers, the procedure is currently not optimized for practical applications. The effectiveness of procedures based on applications of large plastic deformations consists mainly in a distinct decrease of grain size. There is also an obvious benefit in controlled textures of deformed SMA, which can also be utilized with regard to required final properties.

Although studies monitoring behaviour of multiphase Ti-Ni-X based systems exist, a generally known method for the preparation of SMA is not still available. In the future there should be actions related to preparation of specific alloys with verified behaviour. It is obvious that hand-in-hand with the above mentioned there should also be development focused on the preparation and target modification of properties of these materials.

ACKNOWLEDGEMENT

This paper was created under the project No. CZ.1.05/2.1.00/01.0040 "Regional Materials Science and Technology Centre" within the frame of the operation programme "Research and Development for Innovations" financed by the Structural Funds and from the state budget of the Czech Republic and project GA 106/09/1573 „Optimisation of chemical composition, structural characteristics, mechanical properties of NiTi alloys for bio-mechanical applications".

REFERENCES

1. B. Bertheville, J. E. Bidaux, 2005 Alternative powder metallurgical processing of Ti-rich NiTi shape memory alloys, Scripta Materialia, 52 507512 , 1359-6462

2. N. Bertolino, M. Monagheddu, A. Tacca, P. Giuliani, C. Zanotti, A. U. Tamburini, 2003 Ignition mechanism in combustion synthesis of Ti-Al and Ti-Ni systems, Intermetallics, 11 4146 , 0966-9795

3. V. Brailovski, S. D. Prokoshkin, I. Y. Khmelevskaya, K. E. Inaekyan, V. Demers, S. V. Dobatkin, E. V. Tatyanin, 2006 Structure and properties of the Ti-50.0 at%Ni alloy after strain hardening and nanocrystallizing thermomechanical processing, Materials Transactions, 47 795802 , 1347-5320

4. T. Canta, D. Frunza, D. Sabadus, C. Tintelecan, 1998 Some aspects of energy distribution in rotary forming processes, Journal of Material Processing and Technology, 80-81 , 195201 , 0924-0136

5. S. Choi, K. H. Nam, J. H. Kim, 1997 Upper-bound analysis of the rotary forging of a cylindrical bilet, Journal of Material Processing and Technology, 67 7884 , 0924-0136

6. M. S. Choi, T. Fukuda, T. Kakeshita, 2005 Anomalies in resistivity, magnetic susceptibility and specific heat in iron-doped Ti-Ni shape memory alloys, Scripta Materialia, 53 869873 , 1359-6462

7. D. P. Dautovich, G. R. Purdy, 1965 Phase Transformations in TiNi, Canadian Metallurgical Quarterly, 4 129134 , 1879-1395

8. K. Dehghani, A. A. Khamei, 2010 Hot deformation behavior of 60Nitinol (Ni60 wt%- Ti40 wt.%) alloy: experimental and computational studies, Materials Science and Engineering A, 527 684690 , 0092-1509

9. V. Dembovský, 1978 Plasma metallurgy, SNTL, Praha (in Czech) 0440778

10. V. Dembovský, 1985 Plasma metallurgy: The principles, SNTL, Praha, 0-44499-603-6

11. Y. Du, J. C. Schuster, 1998 Experimental investigation and thermodynamic modeling of the Ni-Ti-C system, Zeitschrift fur Metallkunde, 89 399410 , 1862-5282

12. T. W. Duerig, A. Pelton, D. Stöckel, 1999 An overview of nitinol medical applications, Materials Science and Engineering A, 273-275 , 149160 , 0921-5093

13. J. Frenzel, Z. Zhang, K. Neuking, G. Eggeler, 2004 High quality vacuum induction melting of small quantities of NiTi shape memory alloys in graphite crucibles, Journal of Alloys and Compounds, 385 214221 , 0925-8388

14. J. Frenzel, Z. Zhang, Ch. Somsen, K. Neuking, G. Eggeler, 2007a Influence of carbon on martensitic phase transformations in NiTi shape memory alloys, Acta Materialia 55 13311339 . 1359-6454

15. J. Frenzel, J. Pfetzing, K. Neuking, G. Eggeler, 2007b On the influence of thermomechanical treatments on the microstructure and phase transformation behavior of Ni-Ti-Fe shape memory alloys, Materials Science and Engineering A, 481-482 , 635642 , 0092-1509

16. C. Frick, A. Ortega, J. Tyber, K. Gall, H. Maier, 2004 Multiscale structure and properties of cast and deformation processed polycrystalline NiTi shape-memory alloys, Metallurgical and Material Transactions A, 35 20132021 , 1073-5623

17. F. J. Gili, E. Solano, J. Penal, E. Engel, A. Mendoza, J. A. Planell, 2004 Microstructural, mechanical and cytotoxicity evaluation of different NiTi and NiTiCu shape memory alloys, Journal of Material Science, 15 11811189 , 0022-2461

18. R. Graham, 2004 Characteristics of high purity nitinol in Shape Memory and Superelastic Technologies. Proceedings of conferece SMST Society Inc., Denver, 16-19.4.2004, 712

19. A. M. Hindelah, R. Hosemann, 1988 Paracrystals representing the physical state of matter, Journal of Physics C: Solid State Physics 21 41554162 , 0022-3719

20. J. Jung, G. Ghosh, D. Isheim, G. B. Olson, 2003 Design of nanodispersion strengthened TiNi-base shape memory alloys, Proceedings of the international conference on Shape memory and superelastic technologies- SMST 2003, 096605084550 .

21. I. Khmelevskaya, I. B. Trubitsyna, S. D. Prokoshkin, S. V. Dobatkin, V. V. Stolyarov, E. A. Prokofjev, 2001 Structure and Functional Properties of Ti-Ni-Based Shape Memory Alloys Subjected to Severe Plastic deformation, Nanomaterials by Severe Plastic Deformation, 170176 , Wiley-VCH, Weinheim, Germany, 3-52730-659-5

22. H. Y. Kim, Y. Ikehara, J. I. Kim, H. Hosoda, S. Miyazaki, 2006 Martensitic transformation, shape memory effect and superelasticity of Ti-Nb binary alloys, Acta Materialia, 54 24192425 , 1359-6454

23. R. Kocich, M. Kursa, M. Greger, I. Szurman, 2007 Deformation behavior of shape memory alloys on ECAP process, Acta Metallurgica Slovaca, 13 570576 , 1338-1156

24. R. Kocich, I. Szurman, M. Kursa, J. Fiala, Investigation of influence of preparation and heat treatment on deformation behaviour of the alloy NiTi after ECAE, Materials Science and Engineering A, 512 512 100104 , 0921-5093

25. Kockar, B.; Atli, K.C.; Ma, J.; Haouaoui, M.; Karaman, I.; Nagasako, M.; Kainuma, R. (2010). Role of severe plastic deformation on the cyclic reversibility of a Ti50.3Ni33.7Pd16 high temperature shape memory alloy, Acta Materialia, Vol.58, pp. 6411-6417, ISSN 1359-6454

26. L. Kuchař, J. Drápala, 2000 Metallurgy of pure metals-methods of pure materials purifying, Košice Press. 8-07099-471-1

27. T. Kurita, H. Matsumoto, H. Abe, 2004 Transformation behavior in rolled NiTi, Journal of Alloys and Compounds, 381 158166 , 0925-8388

28. M. Kursa, I. Szurman, J. Drápala, M. Losertová, M. Greger, 2005 Shape memory materials Ni-Ti-Me and possibilities of control of their transformation characteristics, VŠB-TU Ostrava (ed), 8-02480-894-32346

29. H. C. Lin, S. K. Wu, 1994 The tensile behavior of a cold-rolled and reverse-transformed equiatomic TiNi alloy, Acta Metallurgica Materialia, 42 16231631 , 1359-6454

30. Liu, Y. (2003). Mechanical and thermomechanical properties of a Ti50Ni25Cu25 melt spun ribbon, Materials Science and Engineering A, Vol.354, pp. 286-290, ISSN 0921-509

31. X. Liu, Y. Wang, D. Yang, M. Qi, 2008 The effect of ageing treatment on shape-setting and superelasticity of a NiTi stent, Materials Characterization 59 402408 , 1044-5803

32. A. M. Locci, R. Orrù, G. Cao, Z. A. Munir, 2003 Field-activated pressure-assisted synthesis of NiTi, Intermetallics, 11 555571 , 0966-9795

33. Ma, j.l.; Wu, k.h. (2000). Effects of Tantalum addition on transformation behaviour of (Ni51 Ti49)1-x Tax and Ni50 Ti50-y Tay Shape Memory Alloys, Materials Science and Technology, Vol.16, pp. 716-719, ISSN 1743-2847

34. X. L. Meng, W. Cai, L. M. Wang, Y. F. Zheng, L. C. Zhao, L. M. Zhou, 2001 Microstructure of Stress-Induced Martensite in a Ti-Ni-Hf High. Temperature Shape Memory Alloy, Scripta Materialia, 45 11771182 , 1359-6462

35. J. Mentz, M. Bram, H. P. Buchkremer, D. Stöver, 2006 Improvement of Mechanical Properties of Powder Metallurgical NiTi Shape Memory Alloys, Advanced Engineering Materials, 8 247252 , 1527-2648

36. J. Mentz, J. Frenzel, M. F. X. Wagner, K. Neuking, G. Eggeler, H. P. Buchkremer, D. Stöver, 2008 Powder metallurgical processing of NiTi shape memory alloys with elevated transformation temperatures, Materials Science and Engineering A, 491 270278 , 0921-5093

37. M. Morakabati, S. Kheirandish, M. Aboutalebi, A. Karimi-Taheri, S. M. Abbasi, 2010 The effect of Cu addition on the hot deformation behavior of NiTi shape memory Alloys, Journal of Alloys and Compounds, 499 5766 , 0925-8388

38. J. Morgiel, E. Cesari, J. Pons, A. Pasko, J. Dutkiewicz, 2002 Microstructure and martensite transformation in aged Ti-25Ni-25Cu shape memory melt spun ribbons, Journal of Material Science, 37 53195327 , 0022-2461

39. T. Mousavi, F. Karimzadeh, M. H. Abbasi, 2008 Synthesis and characterization of nanocrystalline NiTi intermetallic by mechanical alloying, Materials Science and Engineering A, 487 4651 , 1527-2648

40. T. H. Nam, T. Saburi, K. Shimizu, 1990 Shape memory characteristics associated with the B2- B19 and B19- B19′ transformations in a Ti-40Ni-10Cu (at.%) alloy, Materials Transactions JIM 31 959968 , 1073-5623

41. J. P. Noh, 2001 Phase transformation behaviours and shape memory characteristics of Ti-(45-x)Ni-5Cu-xMo (x=0 3-1 0) alloys, Materials Science and Technology, 17 15441550 , 1743-2847

42. K. Otsuka, C. M. . Wayman, Eds, 1998 Shape Memory Materials, Cambridge University Press, ISBN 052144487X

43. Pacholek, P.; Szurman, I.; Sklenaříková, I.; Kursa, M. (2003). Preparation specificity of shape memory alloys on the base of NiTi, Proceedings of 8th International conference Technology 2003, ISBN 80-227-1935-8, Bratislava, 5. - 7.9. 2003, pp. 38-42

44. S. D. Prokoshkin, I. Y. Khmelevskaya, S. B. Dobatkin, I. B. Trubitsyna, E. V. Tatyanin, V. V. Stolyarov, E. A. Prokofiev, 2005 Alloy composition, deformation temperature, pressure and post-deformation annealing effects in severely deformed Ti-Ni based shape memory alloys, Acta Materialia, 53 27032710 , 1359-6454

45. V. G. Pushin, V. V. Kondratjev, 1994 Processing of nanostructured TiNi-shape memory alloys: Methods, structures, properties, application, Physics of Metals and Metallography, 7 497501 , 1555-6190

46. V. G. Pushin, 2000 Alloys with a Thermomechanical Memory: Structure, Properties, and Application, Physics of Metals and Metallography, S68-S95 , 9096 , 1555-6190

47. V. G. Pushin, R. Z. Valiev, Y. T. Zhu, S. D. Prokoshkin, D. V. Gunderov, L. I. Yurchenko, 2006 Effect of Equal Channel Angular Pressing and Repeated Rolling on Structure, Phase Transformations and Properties of TiNi Shape Memory Alloys, Materials Science Forum, 503-504 , 539540 , 0255-5476

48. K. V. Ramaiah, 2005 Processing of Ni-Ti Shape Memory Alloys, Proceedings of conference Smart materials structures and systems, Bangalore, 4-6.5. 2005, 2330 .

49. S. B. Raz, S. K. Sadrnezhaad, 2004 Effect of VIM frequency on chemical composition, homogeneity and microstructure of NiTi shape memory alloy, Materials Science and Technology, 20 593598 , 1743-2847

50. T. Sakuma, I. Iwata, K. Okita, 2003 Effect of additional elements on erosion and wear characteristics of Ti-Ni shape memory alloy,

Corrosion Engineering, Science and Techology, 52 821832 , 1743-2782

51. L. Schetky, M. H. Wu, 2005 Issues in the further Development of Nitinol Properties and Processing for medical Device Applications, Memry Corporation. 18.10.2010, Available from http//memry.com/technology/pdfs/ASM03_NitinolDevelop.pdf

52. A. V. Sergueeva, C. Song, R. Z. Valiev, A. K. Mukherjee, 2003 Structure and properties of amorphous and nanocrystalline NiTi prepared by severe plastic deformation and annealing, Materials Science and Engineering A, 339 159165 , 0092-1509

53. P. M. Standring, J. R. Moon, E. Appleton, 1980 Plastic deformation produced during indentation phase of rotary forging, Journal of Material Processing and Technology, 7 159167 , 0924-0136

54. H. G. Suzuki, E. Takakura, D. Eylon, 1999 Hot strength and hot ductility of titanium alloys-a challenge for continuous casting process, Materials Science and Engineering A, 263 230236 , 0092-1509

55. Szurman, I.; Kursa, M. (2009). Processing technologies of Ni-Ti based shape memory alloys. Proceedings of 8th European Symposium on Martensitic Transformations, ISBN 978-2-7598-0480-1 Prague, 7-11. 9. 2009, pp. 75-81

56. I. Szurman, M. Kursa, 2010 Methods for Ni-Ti based alloys preparation and their comparison. Proceedings of conference Metal 2010, 978-8-08729-417-8 Rožnov pod Radhoštěm, 18.- 20.5. 2010, 861866

57. W. Tang, M. R. Sandstro, Z. G. Wei, S. Miyazaki, 2000 Experimental investigation and thermodynamic calculation of the TiNiCu shape memory alloys, Metallurgical and Material Transactions A, 31 24232429 , 1073-5623

58. J. Tsai, Ch, R. D. Jean, 1994 Effects of hot working on the martensitic transformation of Ni-Ti alloy, Scripta Metallurgica et Materialia, 30 10271030 , 1359-6462

59. J. V. Van Humbeeck, 2001 Shape memory alloys: A material and technology. Advanced Engineering Materials, 3 837850 , ISSNXXX 1527-2648

60. T. Waitz, V. Kazykhanov, H. P. Karnthaler, 2004 Martensitic phase transformations in nanocrystalline NiTi studied by TEM, Acta Materialia, 52 137143 , 1359-6454

61. T. Waitz, 2005 The self-accommodated morphology of martensite in nanocrystalline NiTi shape memory alloys, Acta Materialia, 53 22732280 , 1359-6454

62. G. C. Wang, J. Guan, G. Q. Zhao, 2005 A photo-plastic experimental study on deformation of rotary forging a ring workpiece. Journal of Material Processing and Technology, 169 108115 , 0924-0136

63. M. H. Wu, 2001 Fabrication of Nitinol materials and components, Proceedings of conference SMST-SMM 2001, 978-0-87849-896-3 285291

64. M. H. Wu, 2002 Fabrication of Nitinol Materials and Components, Materials Science Forum, 394-395 , 284290 , 0255-5476

65. M. Zhang, 1984 Calculating force and energy during rotary forging, in proceedings of conference on rotary metal working processes, Journal of Material Processing and Technology, 7 115124 , 0924-0136

66. Z. Zhang, J. Frenzel, K. Neuking, G. Eggeler, 2005 On the reaction between NiTi melts and crucible graphite during vacuum induction melting of NiTi shape memory alloys, Acta Materiala, 53 39713978 , 1359-6454

67. Z. Zhang, J. Frenzel, K. Neuking, G. Eggeler, 2006 Vacuum induction melting of ternary NiTiX (X = Cu, De, Hf, Zr) shape memory alloys using graphite crucibles, Materials Transactions A, 47 661665 , 1073-5623

68. S. L. Zhu, X. J. Yang, D. H. Fu, L. Y. Zhang, C. Y. Li, Z. D. Cui, 2005 Stress-strain behavior of porous NiTi alloys prepared by powders sintering, Materials Science and Engineering A, 408 264268 , 0092-1509

2

CO AND IN DOPED NI-MN-GA MAGNETIC SHAPE MEMORY ALLOYS: A THOROUGH STRUCTURAL, MAGNETIC AND MAGNETOCALORIC STUDY

Simone Fabbrici [1,2], Giacomo Porcari [2,3], Francesco Cugini [3], Massimo Solzi [3], Jiri Kamarad [4], Zdenek Arnold [4], Riccardo Cabassi [2] and Franca Albertini [2,*]

[1]Laboratory of Micro and Submicro Enabling Technologies for Emilia-Romagna (MIST E-R), via Gobetti 101, I-40129 Bologna, Italy

[2]Institute of Materials for Electronics and Magnetism, National Research Council (IMEM-CNR), Parco Area delle Scienze 37/A, I-43124 Parma, Italy

[3]Dipartimento di Fisica e Scienze della Terra, Università di Parma, Parco Area delle Scienze 7/A, I-43124 Parma, Italy

[4]Institute of Physics AS CR, Na Slovance 2, 182 21 Praha 8, Czech Republic

ABSTRACT

In Ni-Mn-Ga ferromagnetic shape memory alloys, Co-doping plays a major role in determining a peculiar phase diagram where, besides a change in the critical temperatures, a change of number, order and nature of phase transitions (e.g., from ferromagnetic to paramagnetic or from paramagnetic to ferromagnetic, on heating) can be obtained, together with a change in the giant magnetocaloric effect from direct to inverse. Here we present a thorough study of the intrinsic magnetic and structural properties, including their dependence on hydrostatic pressurethat are at the basis of the multifunctional behavior of Co and In-doped alloys. We study in depth their magnetocaloric properties, taking advantage of complementary calorimetric and magnetic techniques, and show that if a proper measurement protocol is adopted they all merge to the same values, even in case of first order transitions. A simplified model for the estimation of the adiabatic temperature change that relies only on indirect measurements is proposed, allowing for the quick and reliable evaluation of the magnetocaloric potentiality of new materials starting from readily available magnetic measurements.

KEYWORDS

magnetic shape memory materials; magnetocaloric effect; multifunctional Heusler alloys

1. INTRODUCTION

In 1996, pioneering work by Ullakko, and O'Handley (and his group) in collaboration with Kokorin introduced a new magnetoelastic effect, the magnetic field induced variant reorientation (MIR) in Ni_2MnGa Heusler alloys [1]. Since then "ferromagnetic shape memory" materials, characterized by the coexistence of martensitic transformation and magnetic order, have become an emerging class of materials where new properties and potential fields of applications have constantly come to light [2]. The strong coupling between magnetic and structural degrees of freedom is at the basis of their extraordinary phenomenology that offers also exciting matter for basic investigation [3]. While MIR up to 12% in the martensitic phase due to a magnetostructural coupling on the mesoscopic scale was mainly found in NiMnGa alloys [4,5], giant properties changes obtained by inducing structural transition by external fields (*i.e.*, magnetic field, pressure, stress) were mainly shown in off-stoichiometric Mn-rich Heuslers composed of different IIIA-VA elements (*i.e.*, In, Sn, Sb) [6]. Magnetic superelasticity and strain recovery [7,8], giant magnetoresistance [9] magnetothermal conductivity [10], magnetocaloric [11–14] and barocaloric [15] effects were demonstrated, making these materials very interesting for multifunctional applications, also allowing the multiple exploitation of external fields. Moreover, exchange bias [16] and ferromagnetic strain glass behavior have been recently evidenced [17].

The ternary phase diagram of the NiMnGa system has been mapped to search for new shape memory alloys and for a systematic study of the relationship between martensitic transformation and Curie temperature [18–20]. It has been shown that martensitic and magnetic transformation temperatures show a dependency of electrons per atom following Hume-Rothery concepts. In, Sn, Sb based families of alloys show a linear dependence of martensitic transformation on valence electrons number [21,22]. It has also been recently shown that chemical order plays a primary role in determining martensitic properties [23,24].

Remarkably, the crystallographic structure of the martensitic phase can be tuned by changing composition from modulated (commensurate and incommensurate) monoclinic 10M and 14M structures to non-modulated tetragonal [25–27]. Structural relationships between lattice parameters in the martensitic phase and austenitic and martensitic lattice parameters determine respectively the maximum possible strain achievable by magnetic field induced variant reorientation and by magnetic field induced transformation (MFIS) in single crystals. In polycrystalline isotropic materials the volume discontinuities between austenite and martensite determine the maximum strain achievable by MFIS. In addition, the discontinuities of lattice parameters and volume enable the possibility to drive the martensitic transformation by external fields (stress and pressure) but influence the thermal hysteresis of martensitic transformation, a detrimental effect for applications.

Modeling the magnetic interactions in martensite and austenite and tuning the magnetization discontinuity (ΔM) at the transformation is a crucial goal to drive the martensitic transformation temperature by magnetic fields and to enable giant magnetic field induced effects, and it has been one of the main goals of our research in this field [20,28–31]. In previous papers we demonstrated that in ternary off-stoichiometric Ga-alloys (both in the case of Ni-rich at expenses of Mn and in Mn-rich at expenses of Ga) it is possible to merge martensitic and Curie temperature in a wide temperature range, obtaining a direct first order transformation from ferromagnetic martensite to paramagnetic austenite on heating [20]. As a consequence ΔM maximization and a much higher magnetocaloric effect (Isothermal entropy change, Δs, increased up to 4-times in Ni-rich composition) were obtained [12].

Interestingly, the Mn-Mn distances in these alloys are close to the limit where the interactions switch from ferromagnetic to antiferromagnetic. That is why they show a variety of magnetic properties depending on even minor changes in the stoichiometry, the atomic order, the lattice parameters and the symmetry of the alloy (whether it is in the high temperature or low temperature phase). We have recently shown that in Mn-rich Ni-Mn-Ga alloys, Co substitution produces important changes in magnetism and structure [32]. Chemical disorder in the Mn-rich Heusler alloys is responsible for competing ferromagnetic and antiferromagnetic interaction (magnetic frustration) because the extra Mn atoms occupy lattice sites of the Ga-sublattice which interact antiferromagnetically with the Mn atoms on the Mn-sublattice because of shorter interatomic Mn-Mn distances [22,33] Co affects differently the magnetic interactions of the two phases in the Mn-rich alloy, strengthening the ferromagnetic interactions in austenite while weakening the ferromagnetic behavior of martensite. As a result, the quaternary alloys show lower Curie temperature for the martensitic phase (T_C^M) than the austenitic one (T_C^A). When the martensitic instability temperature (T_M) is tuned between the two Curie temperatures, an inverse magnetostructural transition, from low moment martensite to high moment austenite, occurs (on heating), characterized by a negative field dependence of the transformation temperature ($dT_M/\mu_0 dH$). Owing to the improved ferromagnetic interactions in austenite and to the increased distance between T_M and T_C^A with respect to the ternary compound, very high values of magnetization discontinuity can be achieved, and also of $|dT_M/\mu_0 dH|$. We showed that a 10-fold increase in magnetization discontinuity and a 6-fold increase in critical temperatures sensitivity can be obtained by substituting up to 9 at% Ni with Co in the parent phase $Ni_{50}Mn_{30}Ga_{20}$ [34].

Thus, a metamagnetic behaviour like In, Sn, Sb-based alloys and inverse magnetocaloric effect has been observed also for Ga alloys, by Co-doping in the Mn-rich stoichiometric region [34]. The high values recently obtained for Co-doped In and Ga-based alloys of magnetocaloric [14,35,36] and barocaloric effect [15,37] make them promising for solid state refrigeration and multifunctional applications, even if important drawbacks, mainly linked to hysteresis and poor mechanical properties, must be overcome.

In the present paper we will examine in depth the effects of Co and In doping on Mn-rich NiMnGa alloys, focusing on four representative samples of the series characterized by different number of first and second order phase transitions of different nature (e.g., from *ferro* to *para* or from *para* to *ferro*, on heating). In the first part of the paper we will present a thorough study of the intrinsic magnetic and structural properties, including their dependence on hydrostatic pressure, that are at the basis of their multifunctional behavior. We will then discuss their magnetocaloric properties, taking advantage of complementary calorimetric and magnetic techniques, and showing that if a proper measurement protocol is adopted they all merge to same values, even in the case of first order transitions.

2. RESULTS AND DISCUSSION

2.1. Critical Temperatures and Magnetic Properties

In this work we have chosen four Co and In-doped Ni_2MnGa samples that provide an excerpt of the various magnetic configurations where the martensitic transformation is realized. Tables 1 and 2 report the critical temperatures and the magnetic and structural figures of interest for the analysis of magnetocaloric effect. In order to minimize the role of temperature in this comparison, the three doped samples were chosen with similar T_M values (Figure 1).

The ternary compound $Ni_{50}Mn_{30}Ga_{20}$ shows coincidence of the magnetic and structural transitions across a ferromagnetic martensite to paramagnetic austenite transformation on heating. The critical temperature is approximately 370 K. Due to the partial antiferromagnetic alignment of the extra Mn atoms in the Mn-rich composition and due to the vicinity to the Curie temperature, the magnetization jump ΔM is negative on heating (Figure 2), being approximately −9.5 $A{\cdot}m^2/Kg$, and the field drag on the transition is quite limited (dT_M/μ_0dH = +0.45 K/T).

A notable effect on the structural stability of the quaternary alloy can also be achieved by partial substitution of the group*p* element, Ga, with homovalent In. The peculiarity of this substitution is the selective effect that can be achieved: in fact, by substituting small amounts of Ga with In (around 2 or 3 at.%), the martensitic transformation temperature is lowered while the Curie temperatures are almost unchanged [29]. The In-doped sample chosen for this review, $Ni_{41}Co_9Mn_{32}Ga_{16}In_2$, shows a so-called paramagnetic gap preceding T_M, due to a substantial shift of T_C

^{M} below both $T_C{}^A$ and T_M. The resulting reverse transformation between a very low moment martensite and a high magnetization austenite allows for maximum values of magnetization discontinuity. In this sample both ΔM and $|dT_M/\mu_0 dH|$ reach the highest values of this series (Table 2), increasing of more than one order of magnitude with respect to the ternary parent sample. ($dT_M/\mu_0 dH$ changing from +0.45 to −5.5 K/T)

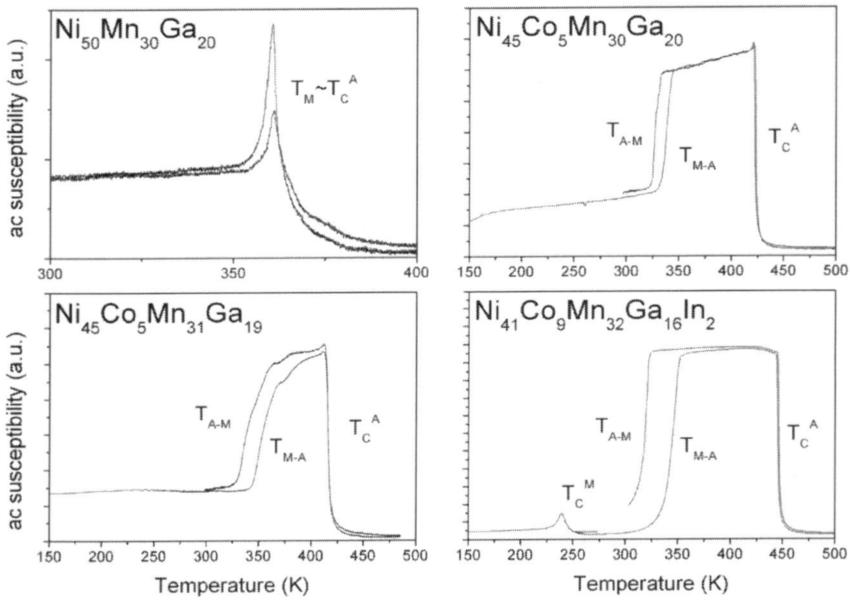

Figure 1. a.c. susceptibility measurements. The nominal compositions and the critical temperatures are highlighted. (Adapted from [29]).

Two quaternary compounds, $Ni_{45}Co_5Mn_{30}Ga_{20}$ (sample 5–30) and $Ni_{45}Co_5Mn_{31}Ga_{19}$ (sample 5–31), display the transformation temperature T_M between two ferromagnetic phases. In fact, by Co doping, the austenitic Curie temperature is pushed to higher values ($T_C{}^A$ = 420 K) while the martensitic transformation temperature is lowered (T_M in the range 330–350 K). Due to enhanced ferromagnetism in austenite, the magnetization jump at the transformation turns to positive values. When compared to the ternary compound, the magnetization discontinuity has reversed sign while field sensitivity of the martensitic transformation is negative and increases three times in absolute value (Table 2 and Figure 2).

Table 1. Critical temperatures, thermal hysteresis ($T_{A\text{-}M}-T_{M\text{-}A}$) and transition width (mean difference between the transformation start and finish temperatures) as measured by a.c. susceptibility data of Figure 1.

Sample ID	Nominal composition	T_C^M [K]	T_C^A [K]	$T_{M\text{-}A}$ [K]	$T_{A\text{-}M}$ [K]	Hysteresis [K]	Transition width [K]
Parent	$Ni_{50}Mn_{30}Ga_{20}$	--	368	375	368	7	10
5-30	$Ni_{40}Co_5Mn_{30}Ga_{20}$	--	420	337	327	10	5
5-31	$Ni_{40}Co_5Mn_{31}Ga_{19}$	--	420	353	341	12	13
In-doped	$Ni_{41}Co_9Mn_{32}Ga_{16}\ In_2$	245	448	343	321	22	35

Table 2. Magnetization variation at the transformation in a 2T field. $\Delta M_{M\text{-}A}$ is calculated as the difference of magnetization at the start and finish temperatures. ΔM_{iso} is the maximum isothermal difference between the heating and cooling branches. magnetic field dependence of T_M. Volume discontinuity at the transition and pressure dependence of T_M. Magnetic field-pressure sensitivity ratio (HPR) and product (HPP) multifunctionality index

Sample ID	$\Delta M_{M\text{-}A}$ [Am^2/Kg]	ΔM_{iso} [Am^2/Kg]	dT_M/μ_0dH [K/T]	$\Delta V/V$ %	dT_M/dp [K/kbar]	HPR $\lvert(dT_M/_0dH)/(dT_M/dp)\rvert$ [kbar/T]	HPP $\lvert(dT_M/_0dH) \times (dT_M/dp)\rvert$ [$K^2/(T\cdot kbar)$]
Parent	−9.5	−7.5	+0.45	0.2	1.0	0.4	0.4
5-30	15	16.5	−1.2	0.45	1.3	0.92	1.56
5-31	16	21.5	−1.3	0.55	2.5	0.52	3.25
In-doped	60	72	−5.5	0.9	6.0	0.91	33

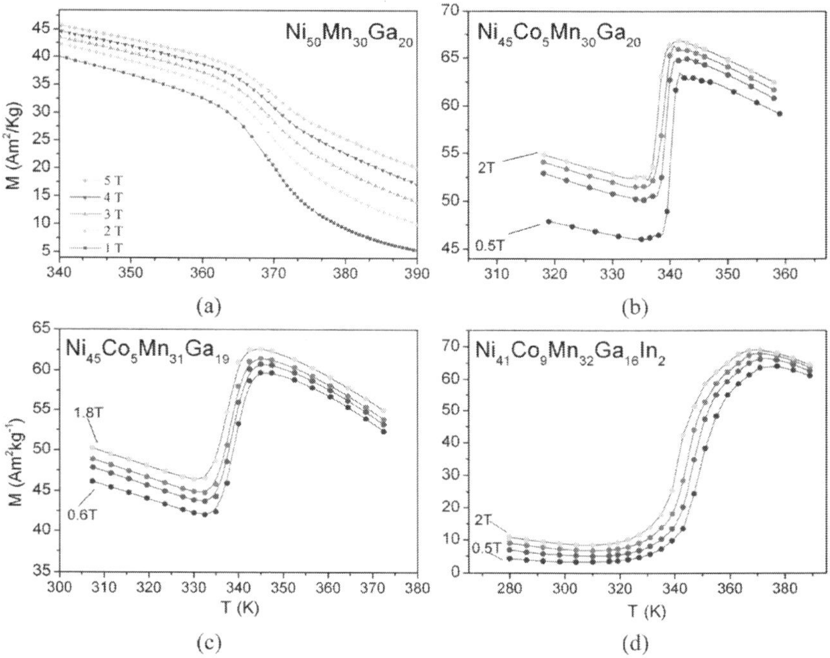

Figure 2. Isofield magnetization measurements *versus* temperature. (**a**) Parent sample; (**b**) 5–30 sample [35]; (**c**) 5–31 sample [36]; (**d**) In-doped sample [35]. The isofield curves are collected at 1T step up to 5T for the parent sample, and at a 0.5T step up to 2T for the others.

2.2. Structural Characterization

In order to get a better understanding of the magnetostructural behavior of these alloys it is necessary to study the structural evolution across the transformation. The samples presented here show that even slight changes in composition result in notable differences. For instance, the two quaternary samples share similar composition, similar magnetic figures (critical temperatures, and dT_M/μ_0dH) and similar hysteresis; nonetheless, sample 5–30 shows a much steeper transformation with respect to sample 5–31 (5 K vs. 13 K transition width). Provided that the two samples share similar dT_M/μ_0dH values, the steeper transition of sample 5–30 allows for complete transformation in a lower applied field. The difference in steepness reflects differently on the magnetization jump at the transition: in fact, the steeper the transition, the more ΔM_{iso} (the maximum achievable ΔM, calculated among the heating and the cooling branches of the transformation) is similar to the transition jump, $\Delta M_{M\text{-}A}$. On the other hand, broader transitions like the one of sample 5–31 show that ΔM_{iso}can be consistently higher than $\Delta M_{M\text{-}A}$.

One of the causes contributing to such a difference may be ascribed to the structural features of the different samples, thus it is interesting to determine the crystal structure of these materials and to follow their evolution across the structural transformation. The crystal structure of the martensitic phase is in both cases tetragonal, non-modulated; nonetheless, the temperature evolution of the X-ray diffraction patterns allows for the determination of different crystal volume discontinuities at the transformation (see for instance Figure 3): sample 5–31 has indeed a higher volume discontinuity than sample 5–30 ($\Delta V/V \approx 0.55\%$ *versus* $\approx 0.45\%$, see Table 2).

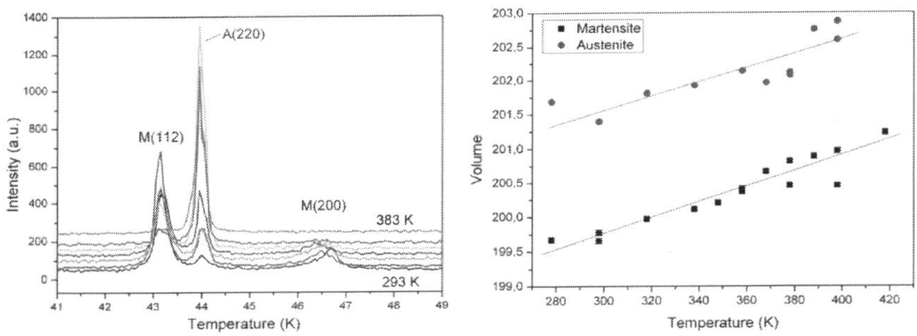

Figure 3. Evolution of the X-ray diffraction pattern with temperature across the transformation for the 5–30 sample (left). Cell Volume evolution of martensite and austenite for the In-doped sample (right) calculated from least square refinement of the diffraction patterns.

The magnetic figures introduced above provide only a partial description of the magnetostructural transformation occurring in these alloys. The martensitic transformation is a first order diffusionless process involving a change in the lattice parameters and in the symmetry of the crystal. Its dynamics depends on the stress accumulated at the border between martensite and austenite, the so-called invariant plane. Symmetry rules and the metric of the two phases, *i.e.*, the mismatch occurring at the invariant plane, [38] determine the entity of the energy barrier which has to be overcome. As we are dealing with polycrystalline alloys, extrinsic contributions to the energy barrier of the transformation arise, too: a complex stress-strain relation among the differently oriented crystallites provides a landscape of energy barriers that heavily influence the width of the martensitic transformation and the thermal hysteresis.

While metric considerations on the lattice matching of the two phases at the invariant plane [38] is a good predictor of the thermal hysteresis, it appears that, at least for series sharing the same martensitic symmetry, the volume discontinuity can be considered a good descriptor of the transformation width; in fact the In doped sample, which show the highest discontinuity in Volume (up to 0.9%), also shows the broader transformation of the series, as reported in Table 2. The parent sample does not follow such a rule; although it shows the lower volume discontinuity of the series ($\Delta V/V \approx 0.2$), its transformation width is comparable with sample 5–31 and even double than sample 5–30. Nonetheless, several features differentiate it from the doped samples, the most important being the different symmetry of the martensitic phase, which in this case is the 7M modulated monoclinic [30]. Also, the presence of the concurrent Curie transition may actually affect the magnetic profile of the transformation. A common feature of all the samples is the sign of the volume discontinuity, which is always positive on heating; this is due to the austenitic phase having always a higher volume than the martensitic one.

The hysteretic nature of the martensitic transformation poses a serious limit to the cycling operation of these alloys [39]. Therefore, the mitigation of cycling irreversibility, either by intrinsic or extrinsic means, is mandatory. The search for new compositions with better martensite-austenite interface matching or the development of proper microstructural configurations are promising routes. A straightforward way to overcome hysteresis and non-negligible transformation width is to rely on values of critical temperature sensitivity to external fields high enough to move the transformation temperature outside of the hysteresis region in moderate fields, taking in mind the bivalent role played by $dT_M/\mu_0 dH$ in determining the magnetocaloric properties [14]. For the sake of applications in energy efficient devices, whether magnetocaloric engines or magneto-thermal energy generators, it is necessary to provide high reversibility in low fields (typically below 2T, the maximum limit of permanent magnets).

On the other hand, the enhanced volume discontinuity that was measured on these alloys allows for improved multifunctionality, *i.e.*, it opens up the possibility of simultaneous shift of the transformation by different fields. Magnetic measurements under hydrostatic pressure were carried out to study the pressure sensitivity of T_M, dT_M/dp; due to increased stability of the most packed structure, martensite, the hydrostatic pressure drives the martensitic temperature to higher values, the opposite of magnetic field application. The dT_M/dp parameters for Co-doped samples, reported in Table 2, are higher than in the ternary compound, and for the In-doped sample reach unprecedented values among the entire class of Ni-Mn based Heuslers [14,29,37]. The In doped sample shows values of dT_M/dp up to 6 K/kbar, 6 times higher than the ternary alloy.

As expected, the trend of the magnetization and volume discontinuities, as well as that of the critical temperature sensitivities to external forces, are coherent, *i.e.*, higher discontinuities provide higher sensitivity of the critical temperature to the corresponding external force (Figure 4).

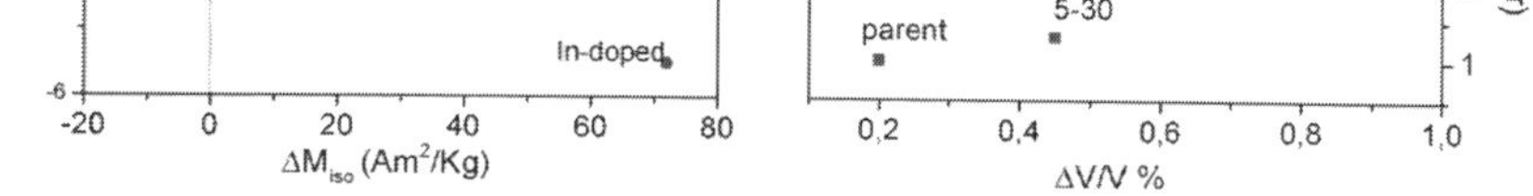

Figure 4. Transformation discontinuity *versus* T_M dependence to the corresponding field: magnetic field (left) and pressure (right).

In our samples the enhanced response to pressure results in a substantial equivalence between 1T and 1 kbar in affecting the transformation temperature. We introduce here a magnetic field-pressure equivalence parameter, HPR = $|(dT_M/\mu_0 dH)/(dT_M/dp)|$, that can be useful to compare the relative field and pressure sensitivity of different alloys (Table 2).

On the other hand, and additional parameter, the product of the two sensitivities, HPP = $|(dT_M/\mu_0 dH)\times(dT_M/dp)|$, can be introduced to account for the increased multifunctionality potential shown by doped alloys (Table 2). By Co and In doping Ni_2MnGa, such multifunctional index is increased by almost two order of magnitudes; this means that both magnetic field and pressure are able to substantially shift the transformation. Such improved sensitivity to both fields, coupled to the opposite effect that magnetic field and pressure display (*i.e.*, magnetic field shifts T_M to lower temperatures, pressure shifts T_M to higher ones) allows for the exploitation of new concept devices where hysteresis can be effectively canceled out by a subsequent application of field and pressure [14]. Figure 5 shows this concept on the In-doped sample: the 5T isofield curves are shifted even outside the thermal hysteresis by the application of 4.5 kbar, effectively overcoming hysteresis.

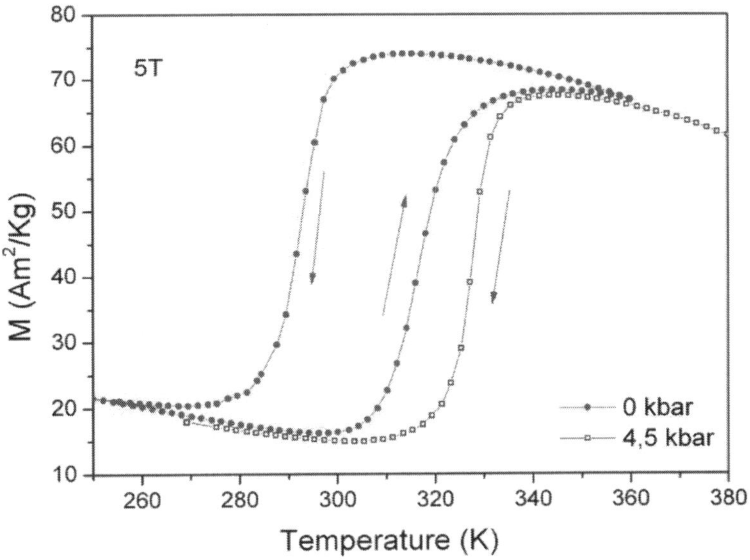

Figure 5. Isofield magnetization curves of the In-doped sample at ambient pressure (full red circles) and 4.5 kbar (open blue squares): the application of 4.5 kbar shifts the cooling branch of the transformation above the heating branch at ambient pressure, outside of the hysteresis region.

2.3. Calorimetry in Magnetic Field

The thermodynamic properties of the presented samples are studied by measuring the temperature and magnetic field dependence of the specific heat across the martensitic transformation (Figure 6). When dealing with magnetocaloric characterization, in-field differential scanning calorimetry (DSC), is indeed the technique which offers the largest amount of information [40]. By comparing the four panels of Figure 6, partial substitution of Ni by Co does not seem to affect the martensitic specific heat baseline ($c_p \sim 500$ J/(Kg·K) at 320 K). The same is not verified for the high temperature phase: the specific heat of austenite in the parent alloy ($Ni_{50}Mn_{30}Ga_{20}$) has lower values than the other materials due to the paramagnetic nature of the high temperature phase.

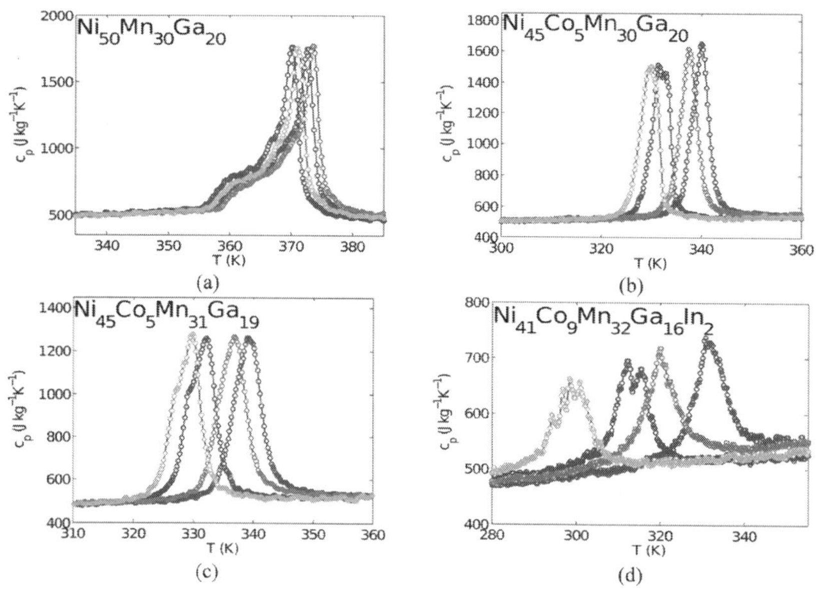

Figure 6. Specific heat curves measured in $\mu_0H = 0$ and 1.8 T. Red and blue curves are heating and cooling in zero field, while orange and cyan are heating and cooling in $\mu_0H = 1.8$ T, respectively. (**a**): parent sample, (**b**): 5–30 sample, (**c**): 5–31 sample [36], (**d**): In-doped sample.

Field dependent calorimetry experiments confirm the values of $dT_M/\mu_0 dH$ and hysteresis width already deduced from magnetization measurements.

It can be noticed that while the direct *ferro-para* system and the Co-doped alloys with *ferro-ferro* inverse transformation show the same $dT_M/\mu_0 dH$ values across the heating and the cooling transformations, the In-doped sample shows unique features. In this case the magnetic field shifts more the cooling branch than the heating one. This behavior, already reported for NiMnSn-based systems [41], is due to the consistently larger ΔM jump across the cooling branch and results in a wider hysteresis in applied field: from 17.7 K in zero field to 20.3 K in 1.8 T. Also, the smaller specific heat peak values reflect a markedly broadened in-field martensitic transformation.

The latent heat λ of the fully transformed transition can be calculated by integrating the c_p profiles after subtraction of the baseline (see Figure 7). Two features stick out from these plots: the value corresponding to the fully transformed process and the field effect on it.

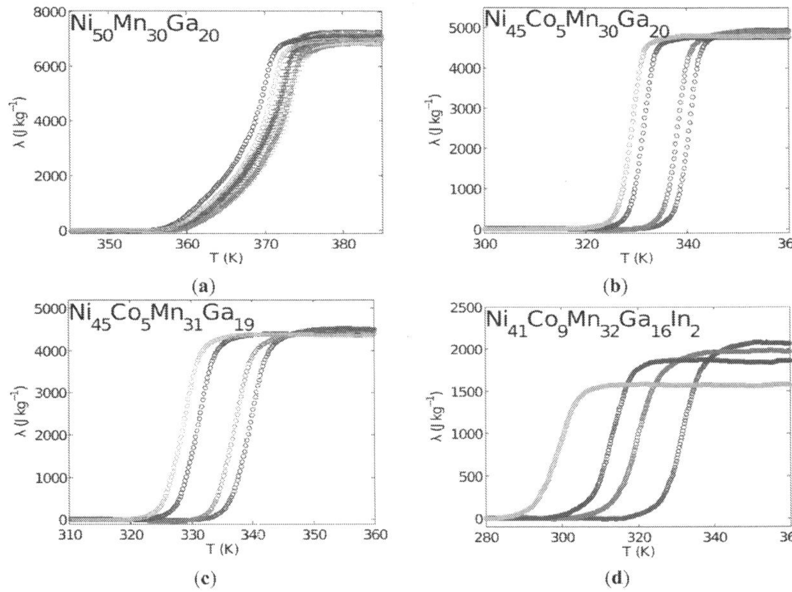

Figure 7. Latent heat out-of-field and in $\mu_0 H$ = 1.8 T, both on heating and cooling. Red and blue are heating and cooling in zero external field, while orange and cyan are heating and cooling in 1.8T. (**a**): parent sample, (**b**): 5–30 sample, (**c**): 5–31 sample, (**d**): In-doped sample.

The parent alloy shows the larger latent heat, pointing out how the effort to increase the MCE by enhancing the volume and the magnetization jump contributes indeed to increase the $|dT_M/\mu_0 dH|$ while the maximum entropy change (calculated as $\Delta S_T = \lambda/T_M$) is depressed. This behavior is striking when comparing the parent alloy with the In-doped one. In the parent sample the ΔS_T of the fully induced process is about −18.5 J/(Kg·K) with $dT_M/\mu_0 dH = +0.45$ K/T, while in the In-doped sample a huge negative $dT_M/\mu_0 dH = -5.5$ K/T (comparable with the systems reported in [14]) can be used to trigger up to 6 J/(Kg·K).

The second interesting aspect is the action of the field on the fully induced latent heat. Within the error bars the direct and the *ferro-ferro* transformations are characterized by a constant transformation latent heat, as measured both on heating and cooling, both in applied and zero field. (Figure 7). The alloy characterized by the "low moment" martensite also in this case displays a peculiar behavior: both for the heating and cooling transformation, the latent heat measured in magnetic field is consistently lower than the corresponding zero field measure. From Figure 7 it can be observed that the stronger decrease of latent heat is realized when the transformation occurs at lower temperature; following the Brillouin-like increase of magnetization of ferromagnetic austenite by lowering the temperature, it appears that the latent heat decreases with ΔM increasing. This phenomenon could be related to a thermodynamic arrest mechanism already pointed out in the literature for Ni-Mn based Heusler alloys showing a reverse magnetostructural transition [42,43]. However, further studies are needed to achieve a better and complete understanding of this feature.

The field dependent specific heat curves are a powerful tool for a comprehensive measurement of the magnetocaloric effect: in fact, by integration it is possible to obtain the entropy-temperature curves across the transformation at different field values. The isentropic and isothermal differences between the two curves represent the adiabatic temperature change, $\Delta T_{ad}(T)$, and isothermal entropy change, $\Delta s_{iso}(T)$, respectively.

Table 3 reports the magnetocaloric data of the samples, both the transition entropy change of the fully induced transformation (calculated from latent heat and magnetic Clausius-Clapeiron relation) and the magnetocaloric effect $\Delta s_{iso}$and ΔT_{ad} induced in a 1.8 T field.

Table 3. Magnetocaloric values of the four samples: Latent heat, transition entropy change calculated from latent heat (Δs_T) and magnetic Clausius-Clapeyron relation ($\Delta s_{C\text{-}C}$); peak values of the isothermal entropy change (Δs_{iso}) and adiabatic temperature change (ΔT_{ad}) in a field span of 1.8 T measured by calorimetric techniques.

Sample	Latent heat (J·kg^{-1})	Δs_T (Jkg^{-1}K^{-1})	$\Delta s_{C\text{-}C}$ (Jkg^{-1}K^{-1})	Δs_{iso}(1.8 T) (Jkg^{-1}K^{-1})	ΔT_{ad}(1.8 T) (K)
$Ni_{50}Mn_{30}Ga_{20}$	6900	−19	−21	−3.7	+0.8
$Ni_{45}Co_5Mn_{30}Ga_{20}$	4900	14	12.5	+6	−1.45
$Ni_{45}Co_5Mn_{31}Ga_{19}$	4500	12.5	12	+5	−1.5
$Ni_{41}Co_9Mn_{32}Ga_{16}In_2$	2300	6	11	+4.5	−2.3

The maximum entropy change calculated from the latent heat, Δs_T, matches quite well the results of magnetic Clausius-Clapeyron equation, $\Delta s_{C\text{-}C}$. The difference shown for the In-doped sample arises from the different pieces of sample used in the different techniques. As reported in [36], proper measurement protocols have to be followed for the reliable analysis of first order transformation materials.

In Figure 8 we compare the temperature behavior of the isothermal entropy change induced by a field span of 1.8 T, Δs_{iso}, calculated by both the magnetic measurements and the field dependent calorimetry. For clarity of the picture, only the heating branch of the transformation is considered. Magnetization isothermal curves have been planned to pick up the full irreversible effect; we took care to cross the cooling martensitic branch before starting every isothermal *M(H)* curve to avoid the presence of "ghost peaks" in the Δs_{iso}(T) curve [44], and calculated the entropy change by exploiting the Maxwell relation. The calorimetric Δs_{iso} values are the isothermal difference between the entropy curves in 0 T and 1.8 T obtained from integration of the specific heat curves of Figure 7. A quite good agreement between the two techniques is observed for all the studied samples (Figure 8). Magnetization measurements and DSC for the Co-substituted alloys have been performed on exactly the same fragment, while the Δs_{iso} curves of the parent $Ni_{50}Mn_{30}Ga_{20}$ sample were obtained from two different samples sharing the same composition: this justifies the slight peak shift (around 3 K) of the two Δs curves.

The maximum values of Δs_{iso} (Table 3) are consistently lower than the ones estimated by latent heat: this demonstrates that the applied field change of 1.8 T is not enough to fully induce the transformation in either of the presented samples. By comparing the four samples, the two *ferro-ferro* inverse alloys, samples 5–30 and 5–31, show the maximum Δs_{iso} of the series. The lower values shown by the parent sample and the In-doped alloy originate from different features: in the first case, the low value of $dT_M/\mu_0 dH$ prevents the applied magnetic field of 1.8 T to fully induce the transformation; in the latter case, both the widening of the transformation and a much lower latent heat concur in obtaining a low Δs_{iso} value.

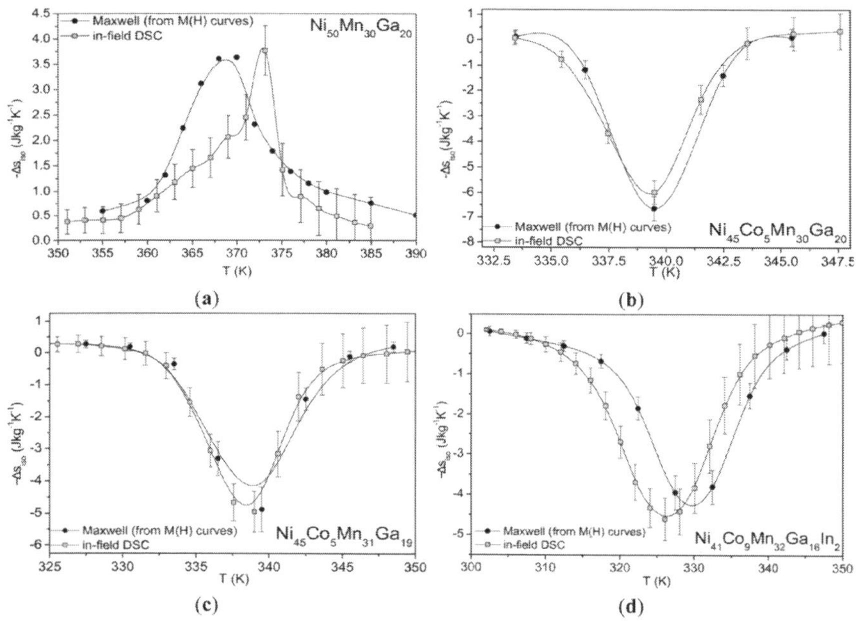

Figure 8. Isothermal entropy change estimated from magnetization isothermal curves (black circles) and field dependent DSC (yellow circles) for $\mu_0\Delta H$ = 1.8 T. (**a**) parent sample; (**b**) 5–30 sample; (**c**) 5–31 sample [36]; (**d**) In-doped sample.

The most interesting question relates at this point to how the large $|dT_M/\mu_0 dH|$ values of the Co substituted alloys can enhance the MCE. The parameter $dT_M/\mu_0 dH$ is well known to set an upper limit to the adiabatic temperature change, while at the same time it gives a rough indication of whether our material may be promising for applications (thanks to the possibility of fully induce the transformation in lower magnetic fields) [45,46].

We noticed that the magnetostructural transformation in these systems can be described, in the frame of a simple geometrical model, by a peculiar relation linking the field induced adiabatic temperature change ΔT_{ad} with $dT_M/\mu_0 dH$, with the martensite specific heat value c_p^{Mart}, the transformation temperature T_M and the isothermal entropy change Δs_{iso}[35,36,47]:

$$\Delta T_{ad} = \frac{\Delta s_{iso} \cdot \Delta T_M}{\Delta s_{iso} + \Delta T_M \cdot \frac{c_p^{Mart}}{T_M}} \tag{1}$$

Here $\Delta T_M = (dT_M/\mu_0 dH) \cdot \mu_0 \Delta H$ is the effective transformation shift in temperature induced by a magnetic field variation $\mu_0 \Delta H$. The usefulness of this equation is that it gives the chance to estimate the ΔT_{ad} merely from indirect magnetization measurements. The specific heat of the low temperature phase can be safely taken from literature, while the remaining parameters can be deduced from isofield magnetization curves. It is important to stress that this relation holds for purely first-order systems and when the field-induced transition shift is smaller than the transformation width [35].

The non-negligible error propagation gives to the calculated ΔT_{ad} an uncertainty of about 30%–40%. The adiabatic temperature change obtained from direct measurements, field dependent DSC and calculated by using the Equation (1) is reported in Figure 9 for samples 5–30 and 5–31. A remarkable convergence of the three experimental techniques is realized: as discussed in [35], these results corroborate the reliability of the presented model for the estimation of the magnetocaloric effect by indirect methods, and highlight how the Maxwell relation for the calculation of isothermal entropy change Δs_{iso}can correctly describe the magnetocaloric effect even of first order transformations. The ΔT_{ad} of these systems seems to increase in proportion to Co content, which means that there is a direct correlation between ΔT_{ad}, ΔM and $dT_M/\mu_0 dH$.

Equation (1) gives also interesting information about the relation linking ΔT_{ad} with $dT_M/\mu_0 dH$ and Δs_{iso}. The first consideration is that ΔT_{ad} is not directly proportional either to $dT_M/\mu_0 dH$ or Δs_{iso}. This is a direct consequence of the fact that the infinitesimal relation $dT_{ad} = (T_M/c_p) ds_{iso}$ cannot be straightforwardly extrapolated to finite differences due to the dependence on magnetic field and temperature of the specific heat, $c_p(H,T)$. On the other hand, eq. 1 relies on the simple value of the specific heat far from the transition region, c_p^{Mart}. Second, and perhaps even more interesting, it can be appreciated that the adiabatic temperature change tends to its upper limit set by the parameter $dT_M/\mu_0 dH$ when $c_p^{Mart}/T_M \rightarrow 0$. This condition is approached for relatively low values of the specific heat of the low temperature phase, and sets an additional figure of merit for the search of promising magnetocaloric materials endowed with high adiabatic temperature change [48].

3. EXPERIMENTAL SECTION

The samples were grown by arc melting stoichiometric amounts of pure elements in an inert atmosphere; homogenization was obtained by subsequent annealing in an inert atmosphere at 1173 K for 72 h followed by water quenching. The final compositions of the grown samples were verified by energy dispersive spectroscopy and were close to the nominal values with a maximum deviation of <1%. The magnetic and magnetostructural critical temperatures were identified by temperature dependence measurements of the initial a.c. susceptibility. The isothermal and isofield magnetic properties were measured in SQUID magnetometers (maximum field 5 T, temperature range 5–500 K).

Magnetic measurements under hydrostatic pressure were performed in a SQUID magnetometer by a purpose-built Cu(Be) pressure cell. Unit cell volumes of the parent and product phases were evaluated by X-ray diffraction at the transformation temperatures using a Thermo ARL X'tra diffractometer equipped with a solid-state Si(Li) Peltier detector and an environmental chamber.

The specific heat measurements, which allowed for the latent heat and magnetocaloric analysis, were performed with purpose built differential scanning calorimeter able to work up to 1.8 T and between 255 K and 390 K. This setup exploits Peltier cells both as active and passive elements [36]. ΔT_{ad} was evaluated for a field change up to 1.9 T at a maximum field rate of <2.2 T/s. The direct probe operates in vacuum (10^{-4} mbar) and uses a Cernox HT-BR temperature sensor connected to the sample with a thermo conductive paste.

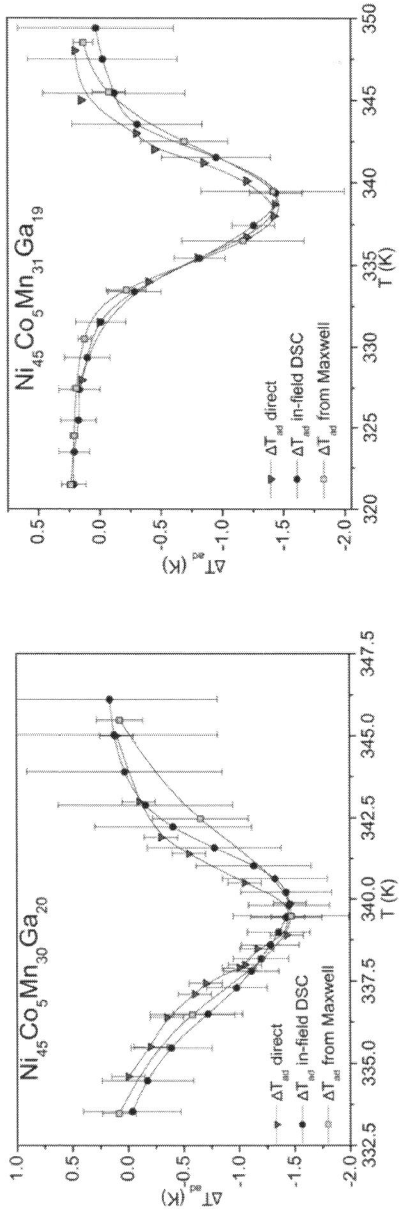

Figure 9. Adiabatic temperature change of samples 5–30 (left) and 5–31 (right) [36] estimated from direct measurements (blue triangles), magnetization data (cyan squares) and field dependent DSC (black circles) in a field span of 1.8 T.

4. CONCLUSIONS

We have shown that in Mn-rich Ni_2MnGa Heusler compounds the chemical substitution with Co and In has profound effects both on the magnetic properties and the structural stability. In particular, it is possible to tune the magnetic properties of austenite and martensite independently, allowing for the martensitic transformation to occur between ferromagnetic phases, or between *ferro*-martensite *para*-austenite or even enabling an inverse transformation between *para*-martensite and a strongly ferromagnetic austenite. This reflects also on the magnetocaloric properties: in this sense the Ga based alloys are attractive for their unique feature to switch from direct to inverse giant MCE by compositional change.

The doped alloys show enhanced magnetic and structural discontinuities across the martensitic transformation, which allow for an increased drive of the structural instability through external stimuli; remarkably, the In-doped sample show the highest value of dT_M/dp (up to 6 K/kbar) for this class of materials.

Since the enhanced response with respect to magnetic field and pressure seems somehow entangled in these alloys, we have suggested two index parameters (the ratio and the product of the critical temperature dependence to external fields) that describe their multifunctionality; such parameters could also be considered for the evaluation and comparison of other classes of materials.

We have reviewed the magneto-thermodynamic properties of the presented samples by several different techniques, providing a comprehensive analysis of the giant magnetocaloric effect across the martensitic transformation. We have proved the convergence of the results obtained from different techniques, and highlighted some detrimental features connected to the magnetostructural transformation, such as inhomogeneity and partial cycling effects, that should be taken care of in order to provide affordable characterization of first order materials. The doped alloys show quite high values of adiabatic temperature change, estimated up to −2.3 K for the In-doped sample.

Finally, we have proposed a simplified model for the estimation of the adiabatic temperature change that relies only on indirect measurements, allowing for the quick and reliable evaluation of the magnetocaloric potentiality of new materials starting from readily available magnetic measurements.

ACKNOWLEDGMENTS

The financial support from the 2007–2013 FESR Operative program of the Emilia Romagna Region (Activity I.1.1) is acknowledged. Part of this work has been performed in the framework of the CNR-AS CR 2009–2012 and 2013–2015 bilateral Agreements.

REFERENCES

1. Ullakko, K.; Huang, J.K.; Kanter, C.; O'Handley, R.C.; Kokorin, V.V. Large magnetic-field-induced strains in Ni_2MnGa single crystals. *Appl. Phys. Lett* **1996**, *69*.

2. Acet, M.; Manosa, L.; Planes, A. Magnetic-Field-Induced Effects in Martensitic Heusler-Based Magnetic Shape Memory Alloys. In *Handbook of Magnetic Materials*; 19Buschow, K.H.J., Ed.; Elsevier: Amsterdam, The Netherlands, 2011; Chapter 4pp. 231–289.

3. Siewert, M.; Gruner, M.E.; Hucht, A.; Herper, H.C.; Dannenberg, A.; Chakrabarti, A.; Singh, N.; Arroyave, R.; Entel, P. A First-Principles Investigation of the Compositional Dependent Properties of Magnetic Shape Memory Heusler Alloys. *Adv. Eng. Mater* **2012**, *14*, 530–546.

4. Sozinov, A.; Likhachev, A.A.; Lanska, N.; Ullakko, K. Giant magnetic-field-induced strain in NiMnGa seven-layered martensitic phase. *Appl. Phys. Lett* **2002**, *80*, 1746–1748.

5. Sozinov, A.; Lanska, N.; Soroka, A.; Zou, W. 12% magnetic field-induced strain in Ni-Mn-Ga-based non-modulated martensite. *Appl. Phys. Lett* **2013**, *102*, 021902.

6. Planes, A.; Manosa, L.; Acet, M. Magnetocaloric effect and its relation to shape-memory properties in ferromagnetic Heusler alloys. *J. Phys. Condens. Matter* **2009**, *21*, 233201.

7. Karaca, H.E.; Karaman, I.; Basaran, B.; Ren, Y.; Chumlyakov, Y.I.; Maier, H.J. Magnetic Field-Induced Phase Transformation in NiMnCoIn Magnetic Shape-Memory Alloys—A New Actuation

Mechanism with Large Work Output. *Adv. Funct. Mater* **2009**, *19*, 983–998.

8. Kainuma, R.; Imano, Y.; Ito, W.; Sutou, Y.; Morito, H.; Okamato, H.; Kitakami, S.; Oikawa, O.; Fujita, A.; Kanomata, T.;*et al.* Magnetic-field-induced shape recovery by reverse phase transformation. *Nature* **2006**, *439*, 957–960.

9. Sharma, V.K.; Chattopadhyay, M.K.; Shaeb, K.H.B.; Chouhan, A.; Roy, S.B. Large magnetoresistance in $Ni_{50}Mn_{34}In_{16}$alloy. *Appl. Phys. Lett* **2006**, *89*, 222509.

10. Zhang, B.; Zhang, X.X.; Yu, S.Y.; Chen, J.L.; Cao, Z.X.; Wu, G.H. Giant magnetothermal conductivity in the Ni-Mn-In ferromagnetic shape memory alloys. *Appl. Phys. Lett* **2007**, *91*, 012510.

11. Hu, F.X.; Shen, B.G.; Sun, J.R. Magnetic entropy change in $Ni_{51.5}Mn_{22.7}Ga_{25.8}$ alloy. *Appl. Phys. Lett* **2000**, *76*, 3460–3462. [

12. Pareti, L.; Solzi, M.; Albertini, F.; Paoluzi, A. Giant entropy change at the co-occurrence of structural and magnetic transitions in the $Ni_{2.19}Mn_{0.81}Ga$ Heusler alloy. *Eur. Phys. J. B* **2003**, *32*, 303–307.

13. Krenke, T.; Duman, E.; Acet, M.; Wassermann, E.F.; Moya, X.; Manosa, L.; Planes, A. Inverse magnetocaloric effect in ferromagnetic Ni-Mn-Sn alloys. *Nat. Mater* **2005**, *4*, 450–454.

14. Liu, J.; Gottschall, T.; Skokov, K.P.; Moore, J.D.; Gutfleisch, O. Giant magnetocaloric effect driven by structural transitions. *Nat. Mater* **2012**, *11*, 620–626.

15. Mañosa, L.; González-Alonso, D.; Planes, A.; Bonnot, E.; Barrio, M.; Tamarit, J.L.; Aksoy, S.; Acet, M. Giant solid-state barocaloric effect in the Ni-Mn-In magnetic shape-memory alloy. *Nat. Mater* **2010**, *9*, 478–481.

16. Khan, M.; Dubenko, I.; Stadler, S.; Ali, N. Exchange bias behavior in Ni-Mn-Sb Heusler alloys. *Appl. Phys. Lett* **2007**, *91*, 072510.

17. Wang, Y.; Huang, C.; Gso, J.; Yang, S.; Ding, X.; Song, X.; Ren, X. Evidence for ferromagnetic strain glass in Ni-Co-Mn-Ga Heusler alloy system. *Appl. Phys. Lett* **2012**, *101*, 101913.

18. Chernenko, V.A.; Cesari, E.; Kokorin, V.V.; Vitenko, I.N. The development of new ferromagnetic shape memory alloys in Ni-Mn-Ga System. *Scripta Metallurgica et Materialia* **1995**, *33*, 1239–1244.

19. Khovaylo, V.V.; Buchelnikov, V.D.; Kainuma, R.; Koledov, V.V.; Ohtsuka, M.; Shavrov, V.G.; Takagi, T.; Taskaev, S.V.; Vasiliev, A.N. Phase transitions in Ni_{2+x} Mn_{1-x} Ga with a high Ni excess. *Phys. Rev. B* **2005**, *72*, 224408.

20. Albertini, F.; Solzi, M.; Paoluzi, A.; Righi, L. Magnetocaloric Properties and Magnetic Anisotropy by Tailoring Phase Transitions in NiMnGa Alloys. In *Advances in Shape Memory Materials*; Book Series: Materials Science Forum; Volume 583, Chernenko, V.A., Ed.; Trans Tech Publications: Zurich, Switzerland, 2008; pp. 169–196.

21. Aksoy, S.; Acet, M.; Wassermann, E.F.; Krenke, T.; Moya, X.; Manosa, L.; Planes, A.; Deen, P.P. Structural properties and magnetic interactions in martensitic Ni-Mn-Sb alloys. *Philos. Mag* **2009**, *89*, 2093–2109.

22. Entel, P.; Dannenberg, A.; Siewert, M.; Herper, H.C.; Gruner, M.E.; Buchelnikov, V.; Chernenko, V.A. Composition-Dependent Basic of Smart Heuslers Materials from First Principles Calculations. In *Advances in Shape Memory Materials*; Book Series: Materials Science Forum; Volume 684, Chernenko, V.A., Ed.; Trans Tech Publications: Zurich, Switzerland, 2011; pp. 1–29.

23. Barandiaran, J.M.; Chernenko, V.A.; Cesari, E.; Salas, D.; Gutierrez, J.; Lazpita, P. Magnetic field and atomic order effect on the martensitic transformation of a metamagnetic alloy. *J. Phys. Condens. Matter* **2013**, *25*, 484005.

24. Entel, P.; Siewert, M.; Grune, M.E.; Herper, H.C.; Comtesse, D.; Arroyave, R.; Singh, N.; Talapatra, A.; Sokolovskiy, V.; Buchelnikov, V.D.; *et al.* Complex magnetic ordering as a driving mechanism of multifunctional properties of Heusler alloys from first principles. *Eur. Phys. J. B* **2013**, *86*, 1–11.

25. Righi, L.; Albertini, F.; Calestani, G.; Pareti, L.; Paoluzi, A.; Ritter, C.; Algarabel, P.A.; Morellon, L.; Ibarra, M.R. Incommensurate modulated structure of the ferromagnetic shape-memory Ni_2MnGa martensite. *J. Solid State Chem***2006**, *179*, 3526–3533.

26. Righi, L.; Albertini, F.; Pareti, L.; Paoluzi, A.; Calestani, G. Commensurate and incommensurate "5M" modulated crystal structures in Ni-Mn-Ga martensitic phases. *Acta Mater* **2007**, *55*, 5237–5245.

27. Righi, L.; Albertini, F.; Villa, E.; Paoluzi, A.; Calestani, G.; Chernenko, V.A.; Besseghini, S.; Ritter, C.; Passaretti, F. Crystal structure of 7M modulated Ni-Mn-Ga martensitic phase. *Acta Mater* **2008**, *56*, 4529–4535.

28. Albertini, F.; Pareti, L.; Paoluzi, A.; Morellon, L.; Algarabel, P.A.; Ibarra, M.R.; Righi, L. Composition and temperature dependence of the magnetocrystalline anisotropy in $Ni_{2+x}Mn_{1+y}Ga_{1+z}$ (x + y + z = 0) Heusler alloys. *Appl. Phys. Lett* **2002**, *81*, 4032–4034.

29. Albertini, F.; Fabbrici, S.; Paoluzi, A.; Kamarad, J.; Arnold, Z.; Righi, L.; Solzi, M.; Porcari, G.; Pernechele, C.; Serrate, D.; *et al.* Reverse Magnetostructural Transitions by Co and In Doping NiMnGa Alloys: Structural, Magnetic, and Magnetoelastic Properties. In *Advances in Magnetic Shape Memory Materials*; Book Series: Materials Science Forum; Volume 684, Chernenko, V.A., Ed.; Trans Tech Publications: Zurich, Switzerland, 2011; pp. 151–163.

30. Righi, L.; Albertini, F.; Fabbrici, S.; Paoluzi, A. Crystal Structures of Modulated Martensitic Phases of FSM Heusler Alloys. In *Advances in Shape Magnetic Memory Materials*; Book Series: Materials Science Forum; Volume 684, Chernenko, V.A., Ed.; Trans Tech Publications: Zurich, Switzerland, 2011; pp. 105–116.

31. Righi, L.; Albertini, F.; Paoluzi, A.; Fabbrici, S.; Villa, E.; Calestani, G.; Besseghini, S. Incommensurate and Commensurate Structural Modulation in Martensitic Phases of FSMA. In *Ferromagnetic Shape Memory Alloys II*; Book Series: Materials Science Forum; Chernenko, V.A., Barandiaran,, J.M., Eds.; Trans Tech Publications: Zurich, Switzerland, 2010; pp. 33–41.

32. Fabbrici, S.; Albertini, F.; Paoluzi, A.; Bolzoni, F.; Cabassi, R.; Solzi, M.; Righi, L.; Calestani, G. Reverse magnetostructural transformation in Co-doped NiMnGa multifunctional alloys. *Appl. Phys. Lett* **2009**, *95*, 022508.

33. Enkovaara, J.; Heczko, O.; Ayuela, A.; Nieminem, R.M. Coexistence of ferromagnetic and antiferromagnetic order in Mn-doped Ni_2MnGa. *Phys. Rev. B* **2003**, *67*, 212405.

34. Fabbrici, S.; Kamarad, J.; Arnold, Z.; Casoli, F.; Paoluzi, A.; Bolzoni, F.; Cabassi, R.; Solzi, M.; Porcari, G.; Pernechele, C.; *et al.* From direct to inverse giant magnetocaloric effect in Co-doped NiMnGa multifunctional alloys. *Acta Mater* **2011**, *59*, 412–419.

35. Porcari, G.; Fabbrici, S.; Pernechele, C.; Albertini, F.; Buzzi, M.; Paoluzi, A.; Kamarad, J.; Arnold, Z.; Solzi, M. Reverse magnetostructural transformation and adiabatic temperature change in Co-, In-substituted Ni-Mn-Ga alloys. *Phys. Rev. B* **2012**, *85*, 024414.

36. Porcari, G.; Cugini, F.; Fabbrici, S.; Pernechele, C.; Albertini, F.; Buzzi, M.; Mangia, M.; Solzi, M. Convergence of direct and indirect methods in the magnetocaloric study of first order transformations: the case of Ni-Co-Mn-Ga Heusler alloys. *Phys. Rev. B* **2012**, *86*, 104432.

37. Manosa, L.; Moya, X.; Planes, A.; Gutfleisch, O.; Lyubina, J.; Barrio, M.; Tamarit, J.L.; Aksoy, S.; Krenke, T.; Acet, M. Effects of hydrostatic pressure on the magnetism and martensitic transition of Ni-Mn-In magnetic superelastic alloys. *Appl. Phys. Lett* **2008**, *92*, 012515.

38. Cui, J.; Chu, Y.S.; Famodu, O.; Furuya, Y.; Hattrick-Simpers, J.; James, R.D.; Ludwig, A.; Thienhaus, S.; Wuttig, M.; Zhang, Z.; *et al.* Combinatorial search of thermoelastic shape-memory alloys with extremely small hysteresis width. *Nat. Mater* **2006**, *5*, 286–290.

39. Emre, B.; Yuce, S.; Stern-Taulats, E.; Planes, A.; Fabbrici, S.; Albertini, F.; Manosa, L. Large reversible entropy change at the inverse magnetocaloric effect in Ni-Co-Mn-Ga-In magnetic shape memory alloys. *J. Appl. Phys* **2013**, *113*, 213905.

40. Basso, V.; Sasso, C.P.; Küpferling, M. A Peltier cells differential calorimeter with kinetic correction for the measurement of c_p(H,T) and Δs(H,T) of magnetocaloric materials. *Rev. Sci. Instrum* **2010**, *81*, 113904.

41. Khovaylo, V.V.; Skokov, K.P.; Gutfleisch, O.; Miki, H.; Takagi, T.; Kanomata, T.; Koledov, V.V.; Shavrov, V.G.; Wang, G.; Palacios, E.; *et al.* Peculiarities of the magnetocaloric properties in Ni-Mn-Sn ferromagnetic shape memory alloys. *Phys. Rev. B* **2010**, *81*, 214406.

42. Ito, W.; Ito, K.; Umetsu, R.Y.; Kainuma, R.; Koyama, K.; Watanabe, K.; Fujita, A.; Oikawa, K.; Ishida, K.; Kanomata, T. Kinetic arrest of

martensitic transformation in the NiCoMnIn metamagnetic shape memory alloy. *Appl. Phys. Lett***2008**, *92*, 021908.

43. Xu, X.; Ito, W.; Umetsu, R.Y.; Koyama, K.; Kainuma, R.; Ishida, K. Kinetic Arrest of Martensitic Transformation in $Ni_{33.0}Co_{13.4}Mn_{39.7}Ga_{13.9}$ Metamagnetic Shape Memory Alloy. *Mat. Trans* **2010**, *51*, 469–471.

44. Caron, L.; Ou, Z.Q.; Nguyen, T.T.; Cam Thanh, D.T.; Tegus, O.; Brück, E. On the determination of the magnetic entropy change in materials with first-order transitions. *J. Magn. Magn. Mater* **2009**, *321*, 3559–3566.

45. Pecharsky, V.K.; Gschneidner, K.A., Jr.; Pecharsky, A.O.; Tishin, A.M. Thermodynamics of the magnetocaloric effect.*Phys. Rev. B* **2001**, *64*, 144406.

46. Skokov, K.P.; Mueller, K.H.; Moore, J.D.; Liu, J.; Karpenkov, A.Y.; Krautz, M.; Gutfleisch, O. Influence of thermal hysteresis and field cycling on the magnetocaloric effect in $LaFe_{11.6}Si_{1.4}$. *J. All. Comp* **2013**, *552*, 310–317.

47. Porcari, G. Magnetocaloric Effect across First Order Transformations of Energy Conversion Materials. Ph.D. Thesis, University of Parma, Parma, Italy, 2013.

48. Pecharsky, V.K.; Gschneidner, K.A., Jr. Some common misconceptions concerning magnetic refrigerant materials. *J. Appl. Phys* **2001**, *90*, 4614–4622.

3

MODELING AND SIMULATION OF SHAPE MEMORY ALLOYS USING MICROPLANE MODEL

Reza Mehrabi[1]

[1] Department of Mechanical Engineering, Vali-e-Asr University of Rafsanjan, Rafsanjan, Iran

ABSTRACT

In this chapter, a three-dimensional phenomenological constitutive model for the simulation of shape memory alloys is introduced. The proposed macromechanical model is based on microplane theory. Microplane approach is chosen to have limited material parameters in that all of those are measurable by simple tests. User material subroutine is developed to implement the proposed model in a commercial finite element package. NiTi hollow tube specimens are under various loading conditions in order to experimentally study the superelastic response of shape memory alloys. Comparing experimental data with numerical results in simple tension and pure torsion as well as proportional and nonproportional tension-torsion loadings demonstrates the capability of proposed model in constitutive modeling of shape memory alloys.

KEYWORDS

Shape memory alloy, phenomenological model, microplane, proportional, nonproportional, experiment

1. INTRODUCTION

Shape memory alloys (SMAs) can recover their original shape when subjected to thermomechanical loading. If the original shape is remembered under external load, it is in superelastic form, and when the original shape is remembered under thermal load, it is in shape memory effect form. Commercial applications of these materials get more attention in recent years. Some of these unique properties are including biocompatibility, good mechanical properties (very similar to the different parts of human body), hysteresis damping, excellent fatigue properties in cyclic loads, and strain hardening. These characteristics make them attractive for diverse fields of application such as biomedical tools (blood clot filters, vascular stents, orthodontic arches), aerospace applications (space shuttle, morphing aircraft, hydraulic fittings and couplings for airplanes), automotive (engines and actuators for smart systems, thermostats), robotic (human robots, artificial muscle), eyeglass frames, cellular phone antennae, coffee maker, and so on [1-9].

In order to design new application of shape memory alloys, basic understanding behavior of these materials under different conditions is essential. Different aspects of SMAs are identified with constitutive modeling. Most models are based on thermodynamics using free energy formulation. However, some models are based on micromechanics concepts or macromechanics model [10]. Micromechanic models deal with crystallographic texture in microscale response of SMAs [11-17]. These models are used to model the phase transformation and grains propagation and consequently are appropriate for modeling the response of SMAs at the microscale [18]. In order to study SMAs polycrystalline, crystallographic texture is a key property. A micromechanic model for single crystals is used to study polycrystal structures [19]. In the micromechanics models, a volume fraction coefficient is defined for variants, and transformation strain is obtained from martensite variants by averaging procedure.

Even though micromechanical models reflect microscopic physical nature of SMAs, they are not suitable for finite element implementation and structural analysis. Therefore, macromechanical models based on phenomenological findings are proposed. These models are efficient in modeling the thermomechanical behavior of SMAs in different engineering applications. In a macromechanic model, macro-scale response of SMAs is considered [20-24]. Phenomenological models belong to a class of macromechanic models that are defined by macroscopic energy functions. These models are calibrated from material characterization and experimental data that depend on the internal variables.

Macrophenomenological models are categorized to one- and three-dimensional models. One-dimensional models are used for the simulation of wire and bar samples under uniaxial loads, while 3-D models can be used for complicated devices under complex loadings. Since most SMA devises are used in multidimensional or under complex loadings, one-dimensional models should extend to three-dimensional models. Three-dimensional constitutive modeling is an important step in analysis and design of SMAs in different industries. Some of the existing macro phenomenological models with specifications and limitations are summarized in Table 1.

Table 1. Macromechanical models

Group	Formulation	Characteristics	Limitation
Tanaka [25]	Helmholtz free energy	(1) 1-D model (2) Exponential phase transformation equation	3-D, shape memory
Liang and Rogers [26]	Helmholtz free energy	(1) 1-D model, extension of the Tanaka model (2) Cosine phase transformation equation	Shape memory
Brinson [27]	Helmholtz free energy	(1) 1-D model, martensite volume fraction separate into two parts induced by stress and temperature (2) Cosine phase transformation equation (3) Different elastic moduli for austenite and martensite	3-D
Auricchio et al. [22, 28]	Based on phase diagram	(1) 1-D and 3-D model (2) Finite-strain regime, rate-dependent behavior (3) Asymmetry behavior and numerical implementation	Shape memory
Lagoudas et al. [2, 20, 29-31]	Gibbs free energy	(1) 3-D model, exponential phase transformation equation (2) Tension-compression asymmetry behavior (3) Nonassociated flow rule during reverse transformation	Depend on the specific alloy
Panico and Brinson [32]	Helmholtz free energy	(1) 3-D model, extension of the Lexcellent model [33, 34] (2) Variant reorientation modeling under nonproportional loadings (3) Computational modeling of porous SMAs	Depend on the specific alloy

Oliveira et al. [36]	Helmholtz free energy	(1) 3-D model (2) Based on Fremond's method (3) Asymmetric behavior and based on plasticity	Unmeasurable material parameter
Brocca et al. [37]	Microplane theory	(1) 3-D model (2) Constant module for different phases (3) Simulation of nonproportional loading	Fundamental concept need to be revised
Arghavani et al. [21, 38]	Helmholtz free energy	(1) 3-D model (2) Finite deformations (3) Martensite volume fraction is a scalar parameter and variant orientation is a tensor	Unmeasurable material parameter
Saleeb et al. [23, 39]	Gibbs free energy	(1) 3-D model (2) Deviation from normality and reorientation under nonproportional loading (3) Cyclic behavior	Unmeasurable material parameter
Mehrabi et al. [40, 41]	Gibbs free energy	(1) 3-D model, microplane formulation	Under investigation
		(2) Anisotropic behavior under nonproportional loadings (3) Tension-compression asymmetry and numerical implementation	
Patoor et al. [15, 39, 42]	Gibbs free energy	(1) 3-D model (2) Tension-compression asymmetry (3) Cyclic behavior and nonproportional loading	Unmeasurable material parameter
Zaki et al. [43-46]	Helmholtz free energy	(1) 3-D model (2) cyclic and asymmetry behavior (3) Capture plastic deformations	Unmeasurable material parameter

One-dimensional constitutive model was proposed by Tanaka [47] based on exponential phase transformation equation. Liang and Rogers [26] proposed a one-dimensional phase transformation based on cosine type. Martensite volume fraction suggested by Liang and Rogers is modified by Brinson [27] into two different parts. These two fractions included martensite volume fraction induced by stress and temperature. This assumption was done to distinct superelastic behavior from shape memory effect. Boyd and Lagoudas [30] had developed 1-D model to 3-D constitutive model. For this development, they used effective stress and strain in phase transformation equations. A three-dimensional constitutive model includes twinned martensite and detwinned martensite. In addition, multidimensional models could predict behavior of shape memory effect as well as superelastic behavior. A 3-D model based on the phase transformation equation of exponential type was proposed by Peng et al. [48]. They used classical plastic theory and defined equivalent stress and strain for modeling SMA response. Reali et al. [22] proposed a 3-D constitutive model, which is capable of simulating superelastic and shape memory behavior. Proposed model is developed within the framework of thermodynamics by defining a scalar and a tensorial internal variable [21]. The 3-D phenomenological model demonstrates the ability of developed model in proportional and nonproportional loadings. In most existing 3-D phenomenological constitutive models, some internal variables are necessary to be calibrated. In calibration process, most models estimate various material parameters in which some of those are not simply measurable by experimental tests. Among various phenomenological models, the microplane model is utilized due to its simplicity and the limited material parameters needed for calibration.

The behavior of some quasi-brittle materials, such as concrete, soil, and stiff foams, is studied using the microplane method [49-52]. A one-dimensional phase transformation model using microsphere formulation was proposed by Ostwald et al. [53] to simulate the polycrystalline materials. The three-dimensional model based on the microplane model was proposed by Brocca et al. [37]. In the microplane model, all material parameters can be determined from uniaxial tension tests at different temperatures. The microplane model considers 1-D equations for some directions on arbitrary plane and is extended to 3-D model using homogenization process [54]. Mehrabi et al. [40, 41, 55] developed this idea within the framework of thermodynamics and proved the capability of the proposed model under multiaxial loadings. Some unique characteristics of SMAs such as deviation from normality in nonproportional loading, anisotropic behavior, and tension-compression asymmetry behavior of SMAs were studied by Mehrabi et al. [56, 57] using the microplane approach.

The proposed microplane model is implemented in a standard commercial finite element package. Some experimental studies under tension, torsion, proportional, and nonproportional loadings have been performed on SMA hollow tubes to assess the proposed model [58]. The numerical results extracted from the microplane model compared with experimental data demonstrate the ability of the microplane approach.

2. PHENOMENOLOGICAL CONSTITUTIVE MODELING

In this section, phenomenological model based on microplane theory is used to describe SMAs material behavior in a simple way. The general definition of the microplane approach is that the 1-D constitutive law is defined for associated normal and tangential stress/strain components on any microplane at each material point. The generalization of 1-D equation to 3-D model is done by homogenization process. In microplane formulation, strain tensor is in closed form of stress tensor. Mehrabi and co-workers have done a thorough research on 3-D phenomenological model based on microplane theory to demonstrate their model features [40, 41, 55, 57, 59, 60]. The three main steps of the microplane model are summarized in Figure 1.

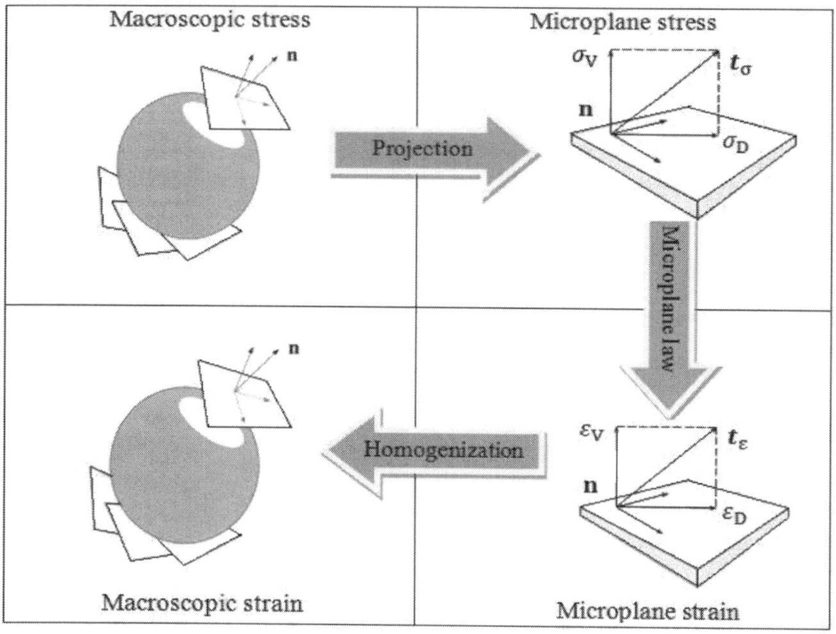

Figure 1. General schematic of the microplane model based on the volumetric-deviatoric split.

For any plane on the shape memory alloys, microscopic Gibbs free energy (G^{mic}) is defined as

$$G^{mic} = \hat{G}^{mic}(\sigma_V, \sigma_D, T, \xi) \tag{1}$$

Here, σV and σDσV and σDare stress components on each microplane, *T* is temperature, and ξξ is martensite volume fraction, which is defined as internal variable [27, 61].

Macroscopic strain tensor is defined as [41]

$$\begin{aligned}\varepsilon &= -\rho\frac{\partial G}{\partial \boldsymbol{\sigma}} = -\rho\frac{3}{2\pi}\int_{\Omega}\frac{\partial G^{mic}}{\partial \boldsymbol{\sigma}}d\Omega = \\ &= \frac{3}{2\pi}\int_{\Omega}\left[-\rho\frac{\partial G^{mic}}{\partial \sigma_V}\frac{\partial \sigma_V}{\partial \boldsymbol{\sigma}} - \rho\frac{\partial G^{mic}}{\partial \boldsymbol{\sigma}_D}\frac{\partial \boldsymbol{\sigma}_D}{\partial \boldsymbol{\sigma}}\right]d\Omega = \\ &= \frac{3}{2\pi}\int_{\Omega}\left(\varepsilon_V\mathbf{V} + \mathbf{Dev}^T.\boldsymbol{\varepsilon}_D\right)d\Omega\end{aligned} \tag{2}$$

These integrals calculated on different orientations of hemisphere at a material point. Projection tensor Dev and the transpose of the deviatoric projection tensor Dev^T are defined as [62]

$$\mathbf{Dev} := \boldsymbol{n}.\mathbf{I}^{dev}, \mathbf{Dev}^T := \mathbf{I}^{dev}.\boldsymbol{n} \tag{3}$$

in which nn represents the unit normal vector on the plane and IdevIdev is deviatoric projection tensor (forth-order identity tensor) that is defined as follows

$$\mathbf{I}^{dev} = \frac{3}{2\pi}\int_{\Omega}\mathbf{Dev}^T.\mathbf{Dev}d\Omega \tag{4}$$

The local 1-D constitutive laws in the volumetric and deviatoric components of the strain for SMAs are defined as

$$\varepsilon_V = \frac{\sigma_V}{E_V^0}, \ \boldsymbol{\varepsilon}_D = \frac{\boldsymbol{\sigma}_D}{E_D^0} + \mathbf{R}\varepsilon^*\xi \tag{5}$$

where ε* is the axial maximum recoverable strain, $E^0_V$and E^0_D are the local linear elastic modulus, which are a function of the global elastic constants. Transformation strain is only initiated in deviatoric direction of microplanes.

Here a standard procedure in the microplane model [57] is used to generalize the 1-D equations to 3-D formulation. Macroscopic strain tensor for shape memory alloys is calculated as

$$\boldsymbol{\varepsilon} = \boldsymbol{\varepsilon}^{e} + \boldsymbol{\varepsilon}^{tr} = \frac{3}{2\pi}\int_{\Omega}\left(\frac{1-2\upsilon}{E(\xi)}\mathbf{V}\otimes\mathbf{V}:\boldsymbol{\sigma} + \frac{1+\upsilon}{E(\xi)}\mathbf{Dev}^{\mathrm{T}}.\mathbf{Dev}:\boldsymbol{\sigma}\right)d\Omega + \frac{3}{2\pi}\int_{\Omega}\varepsilon^{*}\xi\mathbf{Dev}^{\mathrm{T}}.\mathbf{R}\,d\Omega \tag{6}$$

where R is a vector that is defined in reference [41]. In order to numerically calculate these integrals, an integration technique consisting of 21 Gaussian integration points is utilized [63].

3. EXPERIMENTAL STUDY

In order to have a robust constitutive modeling, an experimental study of polycrystalline SMAs gets more attention in recent years. Since some SMA devices experience complex loading paths, investigation of material behavior under multiaxial loadings is essential. Some of the experimental findings are summarized in Table 2. In this table, some of the famous groups and loading schemes on the specific materials are introduced. As it is shown, most experimental studies are on the NiTi materials and copper based SMAs.

A vast experimental study of the NiTi hollow tubes under uniaxial tension, pure torsion, and proportional and nonproportional tension-torsion were done by the author in the Dynamic and Smart Systems Laboratory at the University of Toledo, USA. The Johnson Matthey provided NiTi tube specimens, and the experimental tests were performed using BOSE ElectroForce machine. All mechanical tests were performed at room temperature, and in order to have an isothermal condition, the loading rate was below $10^{-3}s^{-1}$ [76]. The experimental results are compared with numerical findings to show the capability of the proposed approach in the next section.

Table 2. Experimental study

Group	Specification	Material
Sittner et al. [64, 65]	Tension-Torsion	CuAlZnMn
Jacobus et al. [66, 67]	Triaxial stress	CuZnAl, NiTi
Tokuda et al. [68, 69]	Tension-Torsion	CuAlZnMn
Lim and McDowell [70]	Tension-Torsion	NiTi
McNaney et al. [71]	Tension-Torsion	NiTi
Grabe and Bruhns [72, 73]	Tension-Torsion	NiTi
S. Arbab-Chirani and C. Lexcellent [74, 75]	Tension-Pressure-Torsion	CuAlBe, NiTi
Wang et al. [76, 77]	Tension-Torsion	NiTi
Reedlunn et al. [78]	Tension-Pressure-Bending	NiTi
Mehrabi et al. [55-58]	Tension-Torsion	NiTi

4. NUMERICAL SIMULATION

In order to use microplane approach to simulate real SMA devices, the proposed model is implemented and developed into the FE code. The computational algorithm is outlined in Table 3.

Tensile and torsional tests are conducted on the NiTi tubes to investigate the capability of the microplane approach in capturing the behavior of SMAs. The material parameters calibrated for the microplane model are listed in Table 4 [55].

Figures 2 and 3 represent comparison between the axial stress-strain responses of the microplane model and the experimental results as well as shear stress-strain response at room temperature. These comparisons confirm the fact that material parameter calibration process is done as well. Calibrated material parameters are constant during numerical study of proportional and nonproportional loadings.

Table 3. Algorithm for implementation of constitutive modeling of SMA

1. Import the strain increment and the stress evaluated from ABAQUS
2. Check for transformation according to the phase diagram and compute the transformation strain if necessary
3. Compute the elastic strain
4. Compute the total strain
5. Compute the Jacobian matrix
6. Compute the incremental stress tensor
7. Update stress
8. End the program

Table 4. Material properties

Symbols	Values	Units
E_A	20,000	MPa
E_M	13,300	MPa
$\upsilon_A = \upsilon_M$	0.33	
T_f^M	-32	°C
T_s^M	-15	°C
T_s^A	-5	°C
T_f^A	15	°C
σ_s^{cr}	20	MPa
σ_f^{cr}	100	MPa
C_M	6	MPa / °C
C_A	8.2	MPa / °C
ε^*	0.038	

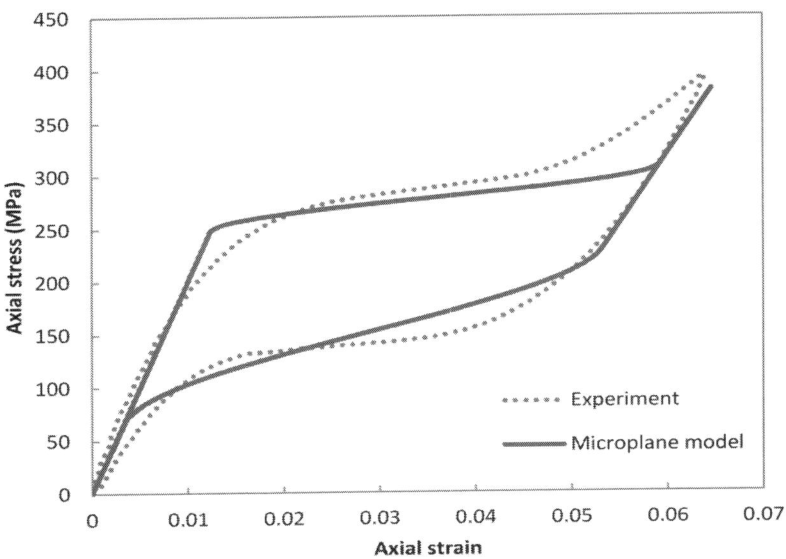

Figure 2. Comparison of the microplane model with experimental result [55].

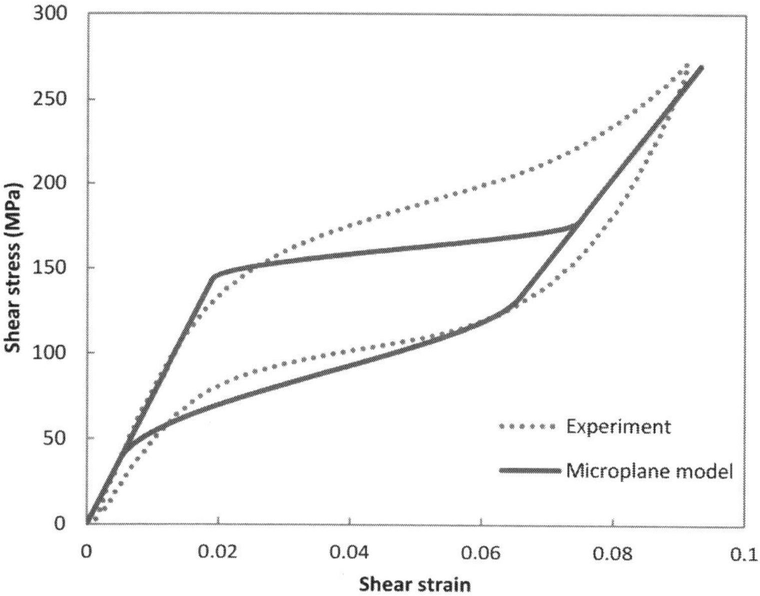

Figure 3. Comparison of the microplane model with experimental result [*55*].

To demonstrate other aspects of the microplane model, proportional tension-torsion loading is experimentally performed. Proportional loading path is shown in Figure 4(a). According to this loading path, axial stress and shear stress are increasing in step 1 and are decreasing to zero during step 2. The experimental findings are compared with the numerical results for axial stress-strain and shear stress-strain in Figures 4(b) and (c). The studied proportional loading demonstrates the capability of the proposed model.

In order to show the capability of the proposed approach in multiaxial loading, one complex loading path is considered here. In Figure 5, nonproportional tension-torsion loading path is shown. At first, shear stress increases while axial stress is zero. During step 2, shear stress is constant, and axial stress increases. Then, shear stress and axial stress are recovered to zero, respectively. Experimental results are compared with microplane numerical results in Figure 6. Comparison of results shows that the proposed model has good agreement with experimental results in both axial stress-strain and shear strain-axial strain. It is obvious that a discrepancy between experimental results and numerical results in the shear stress-strain curve is found. As the proposed model could predict general behavior of SMAs in different loadings, this negligible discrepancy is acceptable.

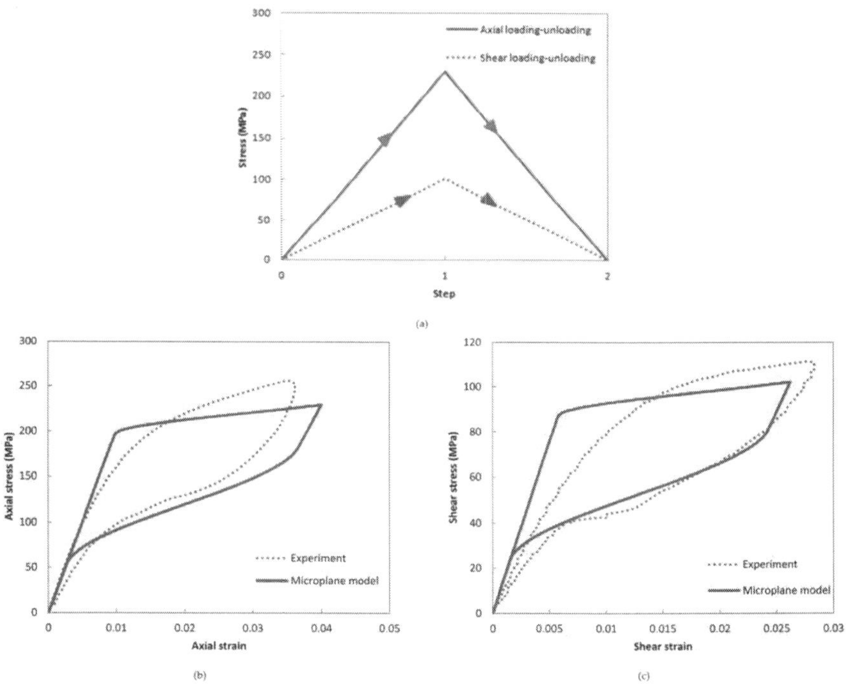

Figure 4. Comparison of the microplane model with experimental result in proportional loading: (a) proportional loading path, (b) axial stress-strain, (c) shear stress-strain [*55*].

In recent years, some biomedical applications [5] such as stent [79], catheters [80], muscle [81], artificial muscles [82], and shape memory implants [83] are produced by SMAs. Therefore, the simulation of biomedical devices using 3-D finite element method [84] is an interesting topic that leads to future works.

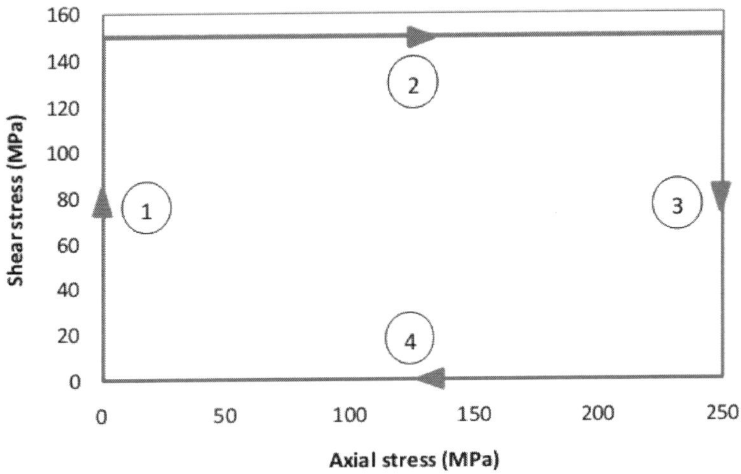

Figure 5. Nonproportional loading path.

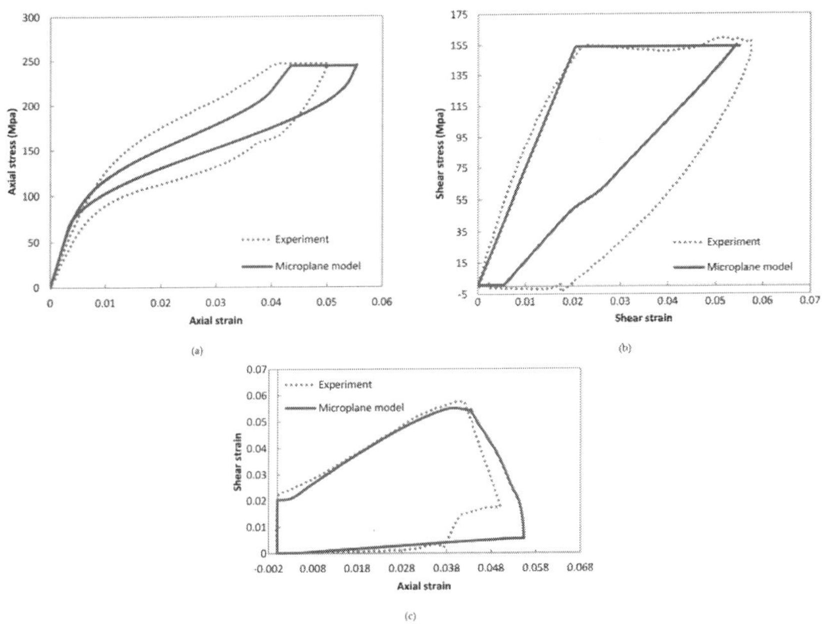

Figure 6. Comparison of the microplane model with experimental result in nonproportional loading: (a) axial stress-strain, (b) shear stress-strain, (c) axial strain-shear strain [*57*].

5. CONCLUSION

Constitutive modeling of shape memory alloys (SMAs) is a key property that leads researchers to find new engineering applications. Phenomenological modeling in macroscopic frame is an appropriate way for modeling the thermomechanical response of SMAs. One of the unique constitutive models based on the microplane model is utilized to investigate behavior of SMAs. Material parameters defined in the proposed model are limited and are calibrated with simple experimental tests. The proposed model is developed to implement and analyze in a finite element package. Some multiaxial loadings as proportional and nonproportional loadings are investigated with constitutive model. Numerical results in comparison of experimental findings show the microplane approach ability in simulation of SMAs behavior.

REFERENCES

1. Anson, T., *Shape memory alloys—medical applications.* Materials World, 1999. 7 (20): p. 745-747.

2. Lagoudas, D.C., *Shape memory alloys: modeling and engineering applications*. 2008: Springer.

3. Lagoudas, D.C. and D.J. Hartl, *Aerospace applications of shape memory alloys.* Proceedings of the Institution of Mechanical Engineers. Part G—Journal of Aerospace Engineering, 2007. 221(G4): p. 535-552.

4. Mohd Jani, J., et al., *A review of shape memory alloy research, applications and opportunities.* Materials and Design, 2014. 56: p. 1078-1113.

5. Petrini, L. and F. Migliavacca, *Biomedical applications of shape memory alloys.* Journal of Metallurgy, 2011. 2011.

6. Song, G., N. Ma, and H.-N. Li, *Applications of shape memory alloys in civil structures.* Engineering Structures, 2006. 28(9): p. 1266-1274.

7. Van Humbeeck, J., *Non-medical applications of shape memory alloys.* Materials Science and Engineering: A, 1999. 273: p. 134-148.

8. Webb, G., et al., *Adaptive control of shape memory alloy actuators for underwater biomimetic applications.* AIAA Journal, 2000. 38(2): p. 325-334.

9. Barbarino, S., et al., *A review on shape memory alloys with applications to morphing aircraft.* Smart Materials and Structures, 2014. 23(6): p. 063001.

10. Brinson, L.C., I. Schmidt, and R. Lammering, *Stress-induced transformation behavior of a polycrystalline NiTi shape memory alloy: micro and macromechanical investigations via in situ optical microscopy.* Journal of the Mechanics and Physics of Solids, 2004. 52(7): p. 1549-1571.

11. Entchev, P.B. and D.C. Lagoudas, *Modeling porous shape memory alloys using micromechanical averaging techniques.* Mechanics of Materials, 2002. 34(1): p. 1-24.

12. Bhattacharya, K. and A. Sadjadpour, *A micromechanics inspired constitutive model for shape-memory alloys: the one-dimensional case.* Smart Materials and Structures, 2007. 16(1): p. S51-S62.

13. Gao, X.J., M.S. Huang, and L.C. Brinson, *A multivariant micromechanical model for SMAs Part 1. Crystallographic issues for single crystal model.* International Journal of Plasticity, 2000. 16(10-11): p. 1345-1369.

14. Levitas, V.I. and I.B. Ozsoy, *Micromechanical modeling of stress-induced phase transformations. Part 1. Thermodynamics and kinetics of coupled interface propagation and reorientation.* International Journal of Plasticity, 2009. 25(2): p. 239-280.

15. Patoor, E., A. Eberhardt, and M. Berveiller, *Micromechanical modelling of superelasticity in shape memory alloys.* Journal De Physique IV, 1996. 6(C1): p. 277-292.

16. Sadjadpour, A. and K. Bhattacharya, *A micromechanics-inspired constitutive model for shape-memory alloys.* Smart Materials and Structures, 2007. 16(5): p. 1751-1765.

17. Thamburaja, P., H. Pan, and F. Chau, *The evolution of microstructure during twinning: constitutive equations, finite-element simulations and experimental verification.* International Journal of Plasticity, 2009. 25(11): p. 2141-2168.

18. Mirzaeifar, R., et al., *A micromechanical analysis of the coupled thermomechanical superelastic response of textured and untextured polycrystalline NiTi shape memory alloys.* Acta Materialia, 2013. 61(12): p. 4542-4558.

19. Thamburaja, P. and L. Anand, *Superelastic behavior in tension-torsion of an initially-textured Ti-Ni shape-memory alloy.* International Journal of Plasticity, 2002. 18(11): p. 1607-1617.

20. Lagoudas, D., et al., *Constitutive model for the numerical analysis of phase transformation in polycrystalline shape memory alloys.* International Journal of Plasticity, 2012. 32: p. 155-183.

21. Arghavani, J., et al., *A 3-D phenomenological constitutive model for shape memory alloys under multiaxial loadings.* International Journal of Plasticity, 2010. 26(7): p. 976-991.

22. Reali, A., F. Auricchio, and U. Stefanelli, *A three-dimensional model describing stress-induced solid phase transformation with permanent inelasticity.* International Journal of Plasticity, 2007. 23(2): p. 207-226.

23. Saleeb, A.F., S.A. Padula, and A. Kumar, *A multi-axial, multimechanism based constitutive model for the comprehensive representation of the evolutionary response of SMAs under general thermomechanical loading conditions.* International Journal of Plasticity, 2011. 27(5): p. 655-687.

24. Sedlák, P., et al., *Thermomechanical model for NiTi-based shape memory alloys including R-phase and material anisotropy under multi-axial loadings.* International Journal of Plasticity, 2012.

25. K. Tanaka, S. Kobayashi, and Y. Sato, *Thermomechanics of transformation pseudoelasticity and shape memory effect in alloys.* International Journal of Plasticity, 1986. 2: p. 59-72.

26. Liang, C. and C. Rogers, *One-dimensional thermomechanical constitutive relations for shape memory materials.* Journal of Intelligent Material Systems and Structures, 1990. 1(2): p. 207-234.

27. Brinson, L.C., *One-dimensional constitutive behavior of shape memory alloys: thermomechanical derivation with non-constant material functions and redefined martensite internal variable.* Journal of Intelligent Material Systems and Structures, 1993. 4(2): p. 229-242.

28. Auricchio, F. and E. Sacco, *A one-dimensional model for superelastic shape-memory alloys with different elastic properties between austenite and martensite.* International Journal of Non-Linear Mechanics, 1997. 32(6): p. 1101-1114.

29. Bekker, A., et al., *Impact induced propagation of phase transformation in a shape memory alloy rod.* International Journal of Plasticity, 2002. 18(11): p. 1447-1479.

30. Boyd, J.G. and D.C. Lagoudas, *Thermomechanical response of shape-memory composites.* Journal of Intelligent Material Systems and Structures, 1994. 5(3): p. 333-346.

31. Hartl, D.J., et al. *Phenomenological modeling of induced transformation anisotropy in shape memory alloy actuators. Proceedings of SPIE.* 2012.

32. Panico, M. and L.C. Brinson, *A three-dimensional phenomenological model for martensite reorientation in shape memory alloys.* Journal of the Mechanics and Physics of Solids, 2007. 55(11): p. 2491-2511.

33. Bouvet, C., S. Calloch, and C. Lexcellent, *A phenomenological model for a Cu-Al-Be shape memory alloy pseudoelasticity under complex loading.* Journal De Physique IV, 2003. 112: p. 201-204.

34. Lexcellent, C. and G. Bourbon, *Thermodynamical model of cyclic behaviour of Ti-Ni and Cu-Zn-Al shape memory alloys under isothermal undulated tensile tests.* Mechanics of Materials, 1996. 24(1): p. 59-73.

35. Helm, D. and P. Haupt, *Shape memory behaviour: modelling within continuum thermomechanics.* International Journal of Solids and Structures, 2003. 40(4): p. 827-849.

36. Oliveira, S., M. Savi, and A. Kalamkarov, *A three-dimensional constitutive model for shape memory alloys.* Archive of Applied Mechanics, 2010. 80(10): p. 1163-1175.

37. Brocca, M., L.C. Brinson, and Z. Bazant, *Three-dimensional constitutive model for shape memory alloys based on microplane model.* Journal of the Mechanics and Physics of Solids, 2002. 50(5): p. 1051-1077.

38. Arghavani, J., et al., *A 3D finite strain phenomenological constitutive model for shape memory alloys considering martensite reorientation.*Continuum Mechanics and Thermodynamics, 2010. 22(5): p. 345-362.

39. Chemisky, Y., et al., *Constitutive model for shape memory alloys including phase transformation, martensitic reorientation and twins accommodation.* Mechanics of Materials, 2011. 43(7): p. 361-376.

40. Mehrabi, R. and M. Kadkhodaei, *3D phenomenological constitutive modeling of shape memory alloys based on microplane theory.* Smart Materials and Structures, 2013. 22(2): p. 025017.

41. Mehrabi, R., M. Kadkhodaei, and M. Elahinia, *A thermodynamically-consistent microplane model for shape memory alloys.* International Journal of Solids and Structures, 2014. 51(14): p. 2666-2675.

42. Lagoudas, D.C., et al., *Shape memory alloys, Part II: modeling of polycrystals.* Mechanics of Materials, 2006. 38(5-6): p. 430-462.

43. Morin, C., Z. Moumni, and W. Zaki, *A constitutive model for shape memory alloys accounting for thermomechanical coupling.* International Journal of Plasticity, 2011. 27(5): p. 748-767.

44. Zaki, W. and Z. Moumni, *A 3D model of the cyclic thermomechanical behavior of shape memory alloys.* Journal of the Mechanics and Physics of Solids, 2007. 55(11): p. 2427-2454.

45. Zaki, W. and Z. Moumni, *A three-dimensional model of the thermomechanical behavior of shape memory alloys.* Journal of the Mechanics and Physics of Solids, 2007. 55(11): p. 2455-2490.

46. Zaki, W., *An efficient implementation for a model of martensite reorientation in martensitic shape memory alloys under multiaxial nonproportional loading.* International Journal of Plasticity, 2012. 37: p. 72-94.

47. Tanaka, K., *Thermomechanical sketch of shape memory effect: one-dimensional tensile behavior.* Res Mechanica: International Journal

of Structural Mechanics and Materials Science, 1986. 18(3): p. 251-263.

48. Peng, X., Y. Yang, and S. Huang, *A comprehensive description for shape memory alloys with a two-phase constitutive model.* International Journal of Solids and Structures, 2001. 38(38-39): p. 6925-6940.

49. Bazant, Z., *Microplane model for strain controlled inelastic behavior.* Mechanics of engineering materials, 1984. 3: p. 45-59.

50. Bažant, Z.P. and B.H. Oh. *microplane model for fracture analysis of concrete structures.* In Proceedings of Symposium on the Interaction of Non-nuclear Munitions with Structures. 1983. Washington.

51. Bažant, Z.P. and P.C. Prat, *Microplane model for brittle plastic material. I: theory.* Journal of Engineering Mechanics, 1988. 114: p. 1672-1688.

52. Carol, I., M. Jirasek, and Z. Bazant, *A thermodynamically consistent approach to microplane theory. Part I. Free energy and consistent microplane stresses.* International Journal of Solids and Structures, 2001. 38(17): p. 2921-2931.

53. Ostwald, R., T. Bartel, and A. Menzel, *A computational micro-sphere model applied to the simulation of phase-transformations.* ZAMM-Journal of Applied Mathematics and Mechanics/Zeitschrift für Angewandte Mathematik und Mechanik, 2010. 90(7-8): p. 605-622.

54. Kadkhodaei, M., et al., *Modeling of shape memory alloys based on microplane theory.* Journal of Intelligent Material Systems and Structures, 2008. 19(5): p. 541-550.

55. Mehrabi, R., et al., *Microplane modeling of shape memory alloy tubes under tension, torsion, and proportional tension-torsion loading.*Journal of Intelligent Material Systems and Structures, 2014: p. 1045389X14522532.

56. Mehrabi, R., et al., *Anisotropic behavior of superelastic NiTi shape memory alloys; an experimental investigation and constitutive modeling.*Mechanics of Materials, 2014. 77: p. 110-124.

57. Mehrabi, R., M. Kadkhodaei, and M. Elahinia, *Constitutive modeling of tension-torsion coupling and tension-compression asymmetry in NiTi shape memory alloys.* Smart Materials and Structures, 2014. 23(7): p. 75021-75035.

58. Mehrabi, R., et al., *Experimental study of NiTi thin-walled tubes under uniaxial tension, torsion, proportional and non-proportional loadings.*Experimental Mechanics, 2015: p. 1-14.

59. Mehrabi, R., M. Kadkhodaei, and S. Arbab Chirani, *Microplane modeling of martensite reorientation in shape memory alloys.* ECCOMAS 2012, Austria, 10-14 September, 2012.

60. Mehrabi, R., M. Kadkhodaie, and M. Elahinia, *Microplane modeling of shape memory alloys in an alternative formulation.* SMASIS2013, Snowbird, Utah, 2013, 2013.

61. Brinson, L.C. and M.S. Huang, *Simplifications and comparisons of shape memory alloy constitutive models.* Journal of Intelligent Material Systems and Structures, 1996. 7(1): p. 108-114.

62. Leukart, M. and E. Ramm, *A comparison of damage models formulated on different material scales.* Computational Materials Science, 2003. 28(3): p. 749-762.

63. Bažant, P. and B. Oh, *Efficient numerical integration on the surface of a sphere.* ZAMM-Journal of Applied Mathematics and Mechanics/Zeitschrift für Angewandte Mathematik und Mechanik, 1986. 66(1): p. 37-49.

64. Sittner, P., Y. Hara, and M. Tokuda, *Experimental-study on the thermoelastic martensitic-transformation in shape-memory alloy polycrystal induced by combined external forces.* Metallurgical and Materials Transactions A. Physical Metallurgy and Materials Science, 1995. 26(11): p. 2923-2935.

65. Sittner, P., et al., *On transformation pathways of general stress controlled thermoelastic martensitic transformation in shape memory alloys.* Journal De Physique IV, 1996. 6(C1): p. 357-366.

66. Gall, K., et al., *Stress-induced martensitic phase transformations in polycrystalline CuZnAl shape memory alloys under different stress states.* Metallurgical and Materials Transactions A, 1998. 29(3): p. 765-773.

67. Jacobus, K., H. Sehitoglu, and M. Balzer, *Effect of stress state on the stress-induced martensitic transformation in polycrystalline Ni-Ti alloy.*Metallurgical and Materials Transactions A. Physical Metallurgy and Materials Science, 1996. 27(10): p. 3066-3073.

68. Tokuda, M., et al., *Multi-axial constitutive equations of polycrystalline shape memory alloy—(Experimental background).* JSME International Journal Series A. Solid Mechanics and Material Engineering, 2002. 45(2): p. 276-281.

69. Tokuda, M., et al., *Thermomechanical behavior of shape memory alloy under complex loading conditions.* International Journal of Plasticity, 1999. 15(2): p. 223-239.

70. Lim, T.J. and D.L. McDowell, *Mechanical behavior of an Ni-Ti shape memory alloy under axial-torsional proportional and nonproportional loading.* Journal of Engineering Materials and Technology-Transactions of the Asme, 1999. 121(1): p. 9-18.

71. McNaney, J.M., et al., *An experimental study of the superelastic effect in a shape-memory Nitinol alloy under biaxial loading.* Mechanics of Materials, 2003. 35(10): p. 969-986.

72. Bruhns, O.T. and C. Grabe, *Path dependence and multiaxial behavior of a polycrystalline NiTi alloy within the pseudoelastic and pseudoplastic temperature regimes.* International Journal of Plasticity, 2009. 25(3): p. 513-545.

73. Grabe, C. and O.T. Bruhns, *Tension/torsion tests of pseudoelastic, polycrystalline NiTi shape memory alloys under temperature control.*Materials Science and Engineering A. Structural Materials Properties Microstructure and Processing, 2008. 481: p. 109-113.

74. Lexcellent, C., et al., *Experimental and numerical determinations of the initial surface of phase transformation under biaxial loading in some polycrystalline shape-memory alloys.* Journal of the Mechanics and Physics of Solids, 2002. 50(12): p. 2717-2735.

75. Saint-Sulpice, L., S.A. Chirani, and S. Calloch, *A 3D super-elastic model for shape memory alloys taking into account progressive strain under cyclic loadings.* Mechanics of Materials, 2009. 41(1): p. 12-26.

76. Wang, X., et al., *An experimental study of the superelastic behavior in NiTi shape memory alloys under biaxial proportional and non-*

proportional cyclic loadings. Mechanics of Materials, 2010. 42(3): p. 365-373.

77. Wang, Y., Z. Yue, and J. Wang, *Experimental and numerical study of the superelastic behaviour on NiTi thin-walled tube under biaxial loading.*Computational Materials Science, 2007. 40(2): p. 246-254.

78. Reedlunn, B., et al., *Tension, compression, and bending of superelastic shape memory alloy tubes.* Journal of the Mechanics and Physics of Solids, 2013.

79. Thériault, P., et al., *Finite element modeling of a progressively expanding shape memory stent.* Journal of biomechanics, 2006. 39(15): p. 2837-2844.

80. Migliavacca, F., et al., *Stainless and shape memory alloy coronary stents: a computational study on the interaction with the vascular wall.*Biomechanics and Modeling in Mechanobiology, 2004. 2(4): p. 205-217.

81. Stirling, L., et al., *Applicability of shape memory alloy wire for an active, soft orthotic.* Journal of Materials Engineering and Performance, 2011. 20(4-5): p. 658-662.

82. Shiraishi, Y., et al. *Development of an artificial myocardium using a covalent shape-memory alloy fiber and its cardiovascular diagnostic response.* In *Engineering in Medicine and Biology Society, 2005. 27th Annual International C*onference of the IEEE-EMBS, 2005. IEEE.

83. Kim, Y.-S., H.-Y. Zhang, and C. Illustrations, Shape Memory Implant. Dynamic Reconstruction of the Spine, 2006: p. 292.

84. Mehrabi, R., KaramoozRavari, M.R., *Superelastic Simulation of SMA Helical Springs.* Smart Structures and Systems, An International Journal, 2015. 16 (1): p.183-194.

4

IN VIVO AND IN VITRO NITINOL CORROSION PROPERTIES

Melissa K. Lonn[1], Justin M. Metcalf[2], Brian D. Choules[2]

[1]Cook Research Incorporated, 1 Geddes Way, West Lafayette, IN 47906, USA

[2]MED Institute, Inc., West Lafayette, IN, USA

ABSTRACT

Regulatory authorities often require in vitro testing on medical devices prior to approval. Current standardized corrosion testing methods (ASTM F2129) require testing in a non-physiologic, de-oxygenated solution for a pre-exposure time of ≤1 h; however, no correlations between the prescribed simulated environment and whole blood conditions have been elucidated. This study compared open circuit potential (OCP), breakdown potentials (E_b), E_b – OCP, and cyclic polarization curves tested in vivo (OCP only) and in vitro in whole blood to those tested in phosphate-buffered saline (PBS). Two oxide thicknesses of Nitinol, two solution oxygen contents (deaerated and aerated solutions), and two pre-exposure durations (acute and chronic) were investigated. The in vitro OCP in whole blood was not significantly different than the in vivo OCP, suggesting that whole blood in vitro can be used to determine baseline corrosion behavior of medical implants. E_b – OCP tested per ASTM F2129 was comparable to acute whole blood and was conservative compared to chronic whole blood for both oxide thicknesses. However, OCP, E_b, and cyclic polarization curves were not always comparable to whole blood. Testing in aerated PBS achieved E_b, E_b – OCP, and cyclic polarization curves that were comparable to or more conservative than whole blood testing, regardless of pre-exposure duration and oxide thickness.

KEYWORDS

NiTi < materials; Shape memory stents; Corrosion; In vivo; Open circuit potential; Breakdown potential

1. INTRODUCTION

In vitro testing is essential for gaining insight into the performance of medical devices. Ideally, in vitro tests closely represent the clinical performance of the device. However, a poorly designed test may cause many hours of unnecessary design iterations, testing, and regulatory activities in an effort to interpret results which do not adequately mimic clinical performance.

It is imperative to fully understand how in vitro results correlate to in vivo results. Designing in vitro tests to mimic chronic physiologic conditions is necessary, and often the tests need to be run in a short amount of time. However, interpretation of short-term results to predict chronic conditions can be difficult. ASTM F2129 is a well accepted in vitro corrosion test for metallic implants that describes a 1–2 h test in a deaerated solution to assess chronic corrosion resistance [1]; however, defining a clinically relevant acceptance criterion has proven difficult and controversial [2]. A clinically relevant acceptance criterion for Nitinol (NiTi) is especially important, since there are multitudes of ways to prepare the surface, which can dramatically affect the ASTM F2129 results.

To develop a clinically relevant acceptance criterion, the open circuit potential (OCP or the rest potential) of the sample in vivo must be understood. Several in vivo studies have been performed on NiTi [3, 4]; however, the reported range of NiTi rest potentials was quite large (−330 mV_{SCE} to +10 mV_{SCE}), and the results were not directly compared to in vitro results.

It has been previously recommended to set an acceptance criterion without consideration of the test sample OCP [5, 6]. Specifically, it has been recommend that $E_b < 300$ mV is unacceptable, E_b between 300 and 600 mV is marginal, and $E_b > 600$ mV is acceptable. It is important to note that these recommendations were made with consideration of only a few metallic alloys (316 stainless steel, cobalt chromium, titanium, and Nitinol). The application of an absolute acceptance criterion for breakdown potential is problematic, since the OCP is different for various metals and metallic surface finishes, and varies with implant duration. Therefore, due to the wide variety of OCPs, an absolute acceptance criterion related to breakdown alone might not be appropriate for the medical device industry.

An alternative to using absolute E_b as the acceptance criterion is to evaluate the corrosion resistance using the difference between breakdown and the open circuit potential (E_b – OCP), which has been recommended previously [7–9]. This difference provides more information about the device's corrosion behavior and how much the device OCP is able to deviate without inducing localized corrosion (i.e., pitting and breakdown). The argument against E_b – E_r (where ASTM F2129 defines E_r, rest potential, as the OCP following ≤1 h deaerated pre-exposure) has been that E_r is measured in a non-physiologic solution that is not representative of in vivo OCP. In vivo OCP values are not well known, and can vary based on immersion solution, oxygen content, immersion time, and surface finish. Several studies have been performed to understand how oxygen content and immersion time affect OCP, E_b, and E_b – OCP [7, 8]; however, no direct correlation with in vivo results has been reported.

Therefore, the aim of this study was to compare the corrosion results of two different NiTi oxide thicknesses tested in vivo (measuring short-term OCP only) and in vitro in whole blood to the results in phosphate-buffered saline (PBS). Additionally, this study investigated the effect of oxygen content of the test solution and pre-exposure time on OCP, E_b, and E_b – OCP.

2. MATERIALS AND METHODS

Testing was conducted on two NiTi oxide thicknesses used for medical devices: black oxide (BO), and electropolished and passivated (EP). These surface finishes were chosen to encompass the range of oxide thicknesses typically used for NiTi medical devices. NiTi #1 (Fort Wayne Metals, Fort Wayne, IN) with a black oxide finish was acquired in the straight annealed condition, with a diameter of 0.026″. NiTi #1 is better than or equal to the chemical composition defined by ASTM F2063. The BO oxide thickness was measured to be 670 Å using Auger electron spectroscopy (Fig. 1a). For the EP processing, BO NiTi samples from the same lot were electropolished and passivated. To illustrate changes in breakdown potential due to testing environment and pre-exposure time, a suboptimal electropolish exhibiting occasional breakdown was chosen for EP so that breakdown would occur in a portion of samples prior to E_v. The resulting oxide thickness was measured to be 34 Å (Fig. 1b). All samples were rinsed with distilled water (ASTM Type 1) and alcohol, and then dried prior to sending to Evans Analytical Group (EAG) for Auger testing and analysis.

Electrochemical corrosion testing was performed with Gamry Reference 600 potentiostats. All potentials were measured in reference to saturated calomel electrodes (SCE). Gamry DC105 software was used to perform the test.

BO and EP NiTi wires were tested in 4 different environments: (1) in vivo arterial blood, (2) in vitro venous whole blood, (3) in vitro aerated PBS, and (4) in vitro deaerated PBS.

2.1 In Vivo Studies

For the in vivo studies, an ovine model was used. Anesthesia was induced, and heparin was administered between 100–200 IU/kg. Testing was approved under facility approved IACUC. Exposed wire (2.1 cm^2) was placed within the animal, while the remainder of the wire was masked off with a non-conductive resin (Miccrostop). The wire was inserted through an 18 French sheath placed within the aorta via femoral access. Bilateral femoral arteries were accessed, allowing two wires to be tested simultaneously. An elongated SCE was also placed within the 18 French sheath, parallel to the test specimen. The tip of the elongated SCE was placed near the wire, outside the sheath. Placement was verified via fluoroscopy. The potential difference between the elongated SCE and the in-house standard SCE was verified to be within ±3 mV. Bilateral BO and EP wires were tested, and the access location was alternated with each new sample to eliminate the potential variable effects of test location and animal. The potential after 1 h was recorded as the OCP. Arterial blood chemistries including pH, partial pressure of oxygen (pO_2), saturation of oxygen (sO_2), hemoglobin (Hb), ionic calcium concentration (iCa), chlorine concentration (Cl), sodium concentration (Na), and potassium concentration (K) were measured before and after each open circuit scan using an Abaxis i-Stat blood analyzer (cartridges CHEM8+ and CG4+).

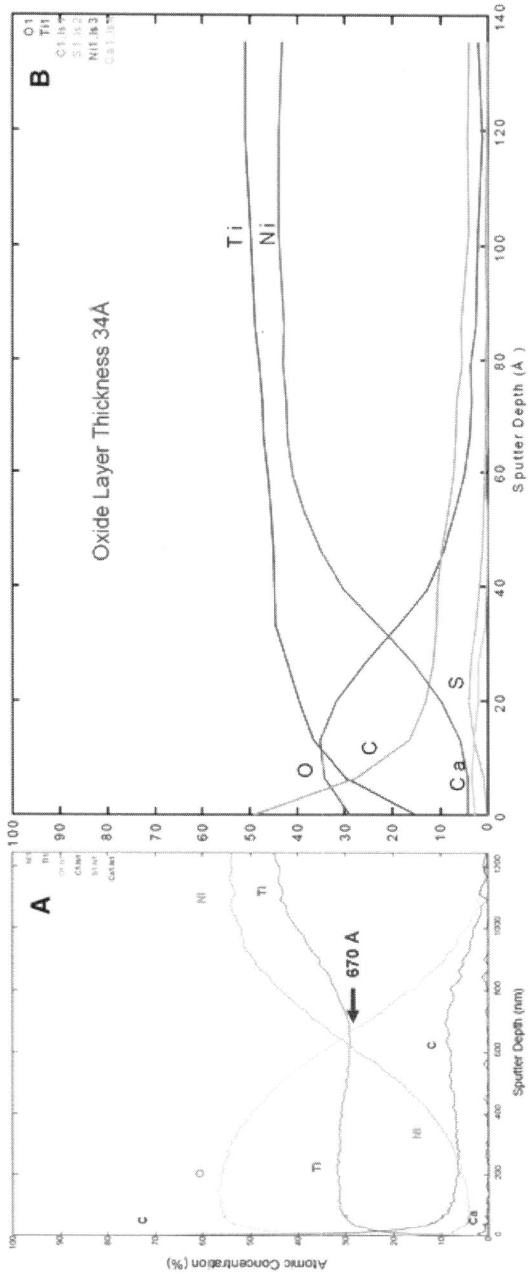

Figure 1. Auger analysis of BO (**a**) and EP (**b**) oxide thicknesses resulting in thicknesses of 670 Å and 34 Å, respectively

2.2 In Vitro Studies

Samples were placed within a 1 L jacketed electrochemical corrosion cell containing whole blood, deaerated PBS, or aerated PBS. Samples were electrically connected to stainless steel mandrels using a silver solder epoxy. Miccrostop was applied to the sample/mandrel interface and the cut ends of the sample. The exposed surface area of the wire samples was 1.6 cm^2. Heparinized bovine whole blood (LAMPIRE Biological Laboratories, Inc., Pipersville, PA) with 1X penicillin/streptomycin was used to prevent bacterial growth. Ionic concentrations were measured from each blood lot. Oxygen concentration, oxygen saturation, and pH were measured before and after each test and daily blood exchanges. PBS [1] was either deaerated (bubbled with nitrogen) or aerated (bubbled with compressed laboratory air) and buffered to a pH of 7.4 ± 0.2. Test solutions were maintained at 37 °C ± 1 °C. Open circuits (or pre-exposures) were measured for either 1 h (acute) or 200 h (chronic), which were monitored discretely for a minimum of 1 h each day until the end of the test. NiTi OCP has been reported to reach steady state after 200 h [7, 8]; therefore, this was chosen to represent a chronic state. Table 1 shows the sample sizes for each OCP test group.

Table 1. Open circuit experimental groups

Surface finish	Solution	Number of samples tested	
		1 h open circuit	200 h open circuit
BO NiTi	In vivo whole blood	8	N/A
	In vitro whole blood	12	4
	In vitro aerated PBS	32	16
	In vitro deaerated PBS	16	8
EP NiTi	In vivo whole blood	8	N/A
	In vitro whole blood	12	4
	In vitro aerated PBS	40	16
	In vitro deaerated PBS	28	16

For all in vitro studies, a cyclic polarization scan was conducted following the open circuit (pre-exposure) delay. Cyclic polarization testing followed ASTM F2129, except for the experimental changes to the aeration and open circuit duration. The forward and reverse scan rates were 1 mV/s. The cyclic polarization scan was started at the OCP, and the E_v (vertex potential) was 1200 mV. Pitting of the samples was confirmed using light microscopy of up to ×25 magnification.

Pre-exposure solution, pre-exposure time, and scan solution were investigated; specifically, samples were assigned to eight different cyclic polarization experimental groups (Table 2). Group 1 (acute) samples and Group 2 (chronic) samples were tested in heparinized bovine whole blood. The natural oxygen content in the whole blood was desired, eliminating the need to bubble the solution. Group 3 (acute) samples and Group 4 (chronic) samples were tested in aerated PBS. Additionally, Group 5 (acute) and Group 6 (chronic) samples were used to investigate the effect of pre-exposure in aerated PBS prior to testing in deaerated PBS, in which case the solution was deaerated for 30 min with nitrogen prior to the scan. Finally, Group 7 (acute—the control group tested fully to ASTM F2129) and Group 8 (chronic) samples were tested in deaerated PBS.

2.3 Data Analysis

Experimental groups were compared using the GLM procedure in SAS (Statistical Analysis Software, version 9.3) with a least-square mean statement utilizing Tukey test ($\alpha = 0.05$). For groups with censored data (i.e., lack of breakdown prior to reversing the scan), a Kaplan–Meier survival curve (lifetest procedure in SAS) was used to compare groups.

3. RESULTS

3.1 Acute and Chronic Open Circuit Potential (OCP)

OCP following 1 h pre-exposure (acute) in solution for all test environments is presented in Fig. 2. The OCP for acute in vitro whole blood was not significantly different from the OCP for in vivo whole blood for both oxide thicknesses ($p = 0.0636$ for BO, 0.9964 for EP). For the acute BO groups, in vitro aerated and deaerated PBS were significantly different from the whole blood groups ($p < 0.0001$). For the acute EP groups, only the aerated PBS group was significantly different from the whole blood groups ($p < 0.0001$). Oxygen content influenced mean OCP for all samples at 1 h (Fig. 3), though the effect was greater on samples with thicker oxides. Therefore, depending on the oxide thickness, testing in conditions performed per ASTM F2129 (deaerated PBS for 1 h) might result in OCPs which are not representative of in vivo OCPs.

OCPs after 200 h (chronic) compared to 1 h (acute) OCPs are shown in Fig. 2. All potentials increased with longer pre-exposure times, and all groups, except EP in deaerated PBS, trended toward a similar potential (11–58 mV). Therefore, even after 200 h, deaerated OCPs were not representative of in vivo potentials for all oxide thicknesses, whereas chronic-aerated OCPs were similar to chronic whole blood OCPs for both oxide thicknesses ($p = 0.9976$ and $p = 0.7397$).

Auger was performed on EP NiTi samples following 1 and 200 h in PBS to compare to the untested samples (Figs. 4 and 5). All samples increased in oxide thickness following testing, with the largest increase occurring in aerated PBS following 200 h. This is consistent with results reported previously, indicating that oxide changes over time and increases corrosion resistance [10, 11].

3.2 Breakdown Potentials (E_b)

The results for E_b of all groups investigated are presented in Fig. 6 using Kaplan–Meier survival curves to account for censored data (i.e., the E_v was reached prior to breakdown). Representative cyclic polarization curves are presented in Fig. 7.

For both oxide thicknesses, E_b improved with longer pre-exposure time. The acute BO samples exposed exclusively to aerated solutions (pre-exposure and scan) were most similar to whole blood ($p = 0.7774$) (Figs. 6 a and 7a). The E_b of chronic BO samples exposed exclusively to deaerated solutions (pre-exposure and scan) and the samples exposed exclusively to aerated solutions were most similar to the whole blood E_b ($p = 0.9850$ and $p = 0.1439$, respectively) (Fig. 6c). Most importantly, the E_b occurring after chronic immersion in whole blood was*significantly* different than the control as tested per ASTM F2129 (1 h deaerated pre-exposure with deaerated scan), $p < 0.0001$. For both acute and chronic EP experimental groups, no significant difference in E_b was observed for the different testing conditions ($p = 0.2297$ and $p = 0.6482$, respectively). However, acute EP samples exposed exclusively to aerated solutions had the most conservative (i.e., resulting in the lowest E_b) breakdown survival curve (Fig. 6b). Overall, fewer EP samples broke down after chronic pre-exposure compared to acute pre-exposure (Fig. 6b and d).

Table 2. Cyclic polarization experimental groups

Cyclic polarization experimental group	Open circuit (OC)		Polarization scan solution	Sample size	
	Solution	Duration (h)		BO NiTi	EP NiTi
1	In vitro whole blood	1	In vitro whole blood	8	8
2		200		4	4
3	In vitro aerated PBS	1	In vitro aerated PBS	8	11
4		200		8	7
5	In vitro aerated PBS	1	In vitro deaerated PBS	8	11
6		200		8	8
7 (control)	In vitro deaerated PBS	1	In vitro deaerated PBS	8	11
8		200		7	7

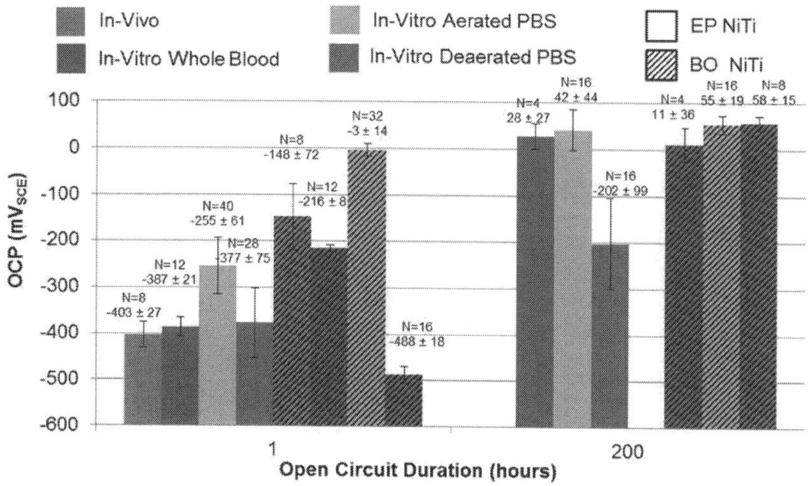

Figure 2. Acute and chronic open circuit potentials

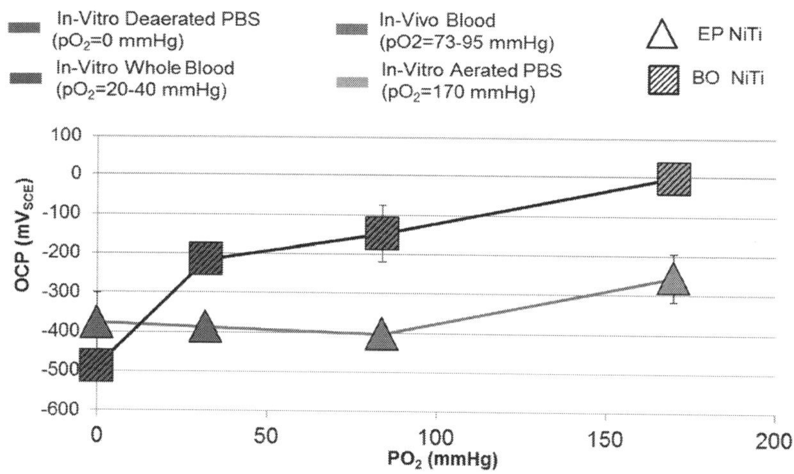

Figure 3. Open circuit potential versus oxygen partial pressure

It is also interesting to note that acute BO samples pre-exposed to aerated solutions followed by a deaerated scan were not significantly different than the samples exposed exclusively to deaerated solutions ($p = 0.9957$, Figs. 6a and 7a), whereas chronic BO samples pre-exposed to aerated solutions followed by a deaerated scan were similar to the samples exposed exclusively to aerated solutions (p = 0.7786, Figs. 6c and 7c).

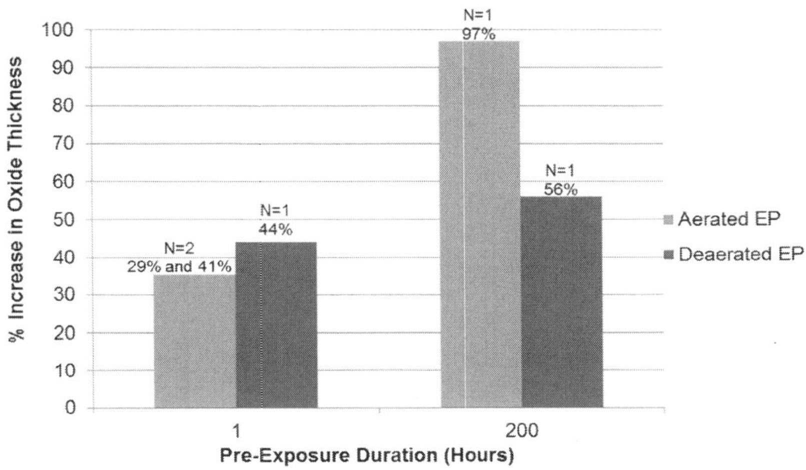

Figure 4. Percent increase in oxide thickness versus pre-exposure duration of EP NiTi wires

3.3 Difference Between E_b and OCP (E_b – OCP)

For both oxide thicknesses, E_b – OCP increased with pre-exposure duration, Fig. 8. For Groups 5 and 6, OCPs were taken after the aerated pre-exposure, prior to deaerating with nitrogen to start the cyclic polarization scan.

The BO samples exposed exclusively to aerated solutions were more conservative (i.e., resulting in a lower E_b – OCP) than whole blood for both acute and chronic time points (Fig. 8 a and c). The E_b – OCP results for BO samples exposed only to deaerated solutions were most similar to whole blood for both acute and chronic durations (Fig. 8 a and c). However, acute cyclic polarization scans for deaerated PBS and whole blood were noticeably different (Fig. 7a). Thus, in vivo conditions are not fully represented with a 1 h deaerated pre-exposure with deaerated scan, depending on oxide thickness.

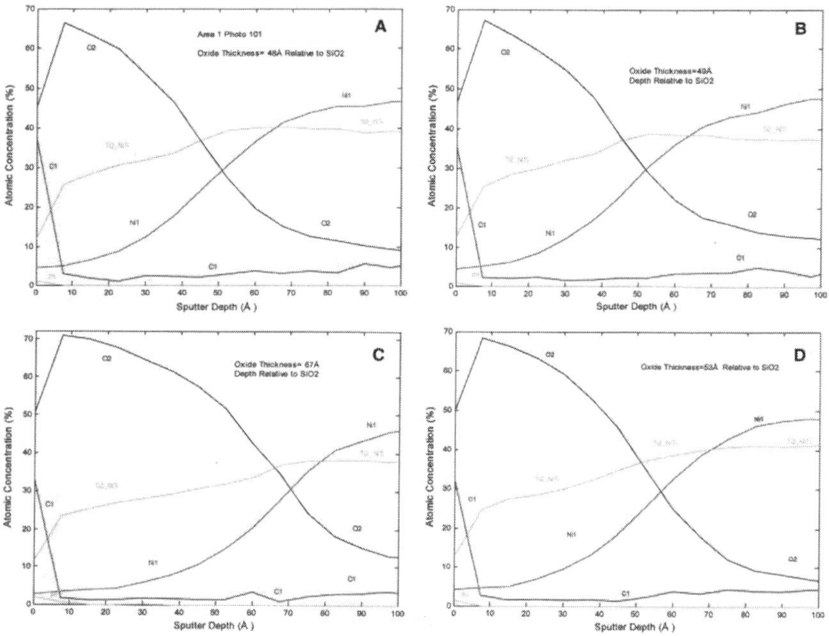

Figure 5. Auger analysis of EP oxide thicknesses following pre-exposure: acute aerated (**a**); acute deaerated (**b**); chronic aerated (**c**); chronic deaerated (**d**)

For both acute and chronic EP experimental groups, no significant difference between the potential difference survival curves was observed for the different testing conditions ($p = 0.0580$ and $p = 0.6482$, respectively). The most conservative condition for acute EP E_b – OCP was the aerated pre-exposure with an aerated scan.

A summary of results is presented in Table 3.

4. DISCUSSION

The cyclic polarization testing standard for metallic devices (ASTM F2129) has been well received by regulatory agencies as an acceptable test for corrosion resistance for implant devices [1]. However, determination of an acceptance criterion has been a topic of much debate. The standard states that the method is intentionally severe and meant for comparison purposes only; however, an absolute acceptance criterion for E_b has been proposed [5, 6]. The standard defines a pre-exposure time ≤1 h; however, oxide thicknesses have been found to change over time [10, 11], which can potentially change the corrosion resistance. Therefore, care must be taken when determining clinically relevant acceptance criteria. Additionally, the standard dictates that testing be conducted in a non-physiological de-oxygenated solution, with the aim of producing a lower E_b. However, blood and other body fluids are not anoxic (typical arterial and venous oxygen partial pressure ranges are 80–100 and 40–50 mmHg, respectively) [12]. Finally, the corrosion standard for galvanic testing of medical implants, ASTM F3044, states that testing should be performed in an aerated solution and that the coupled potential be compared to E_b as found in ASTM F2129; thus, further confounding testing practices when comparing aerated galvanic results to deaerated breakdown results [13].

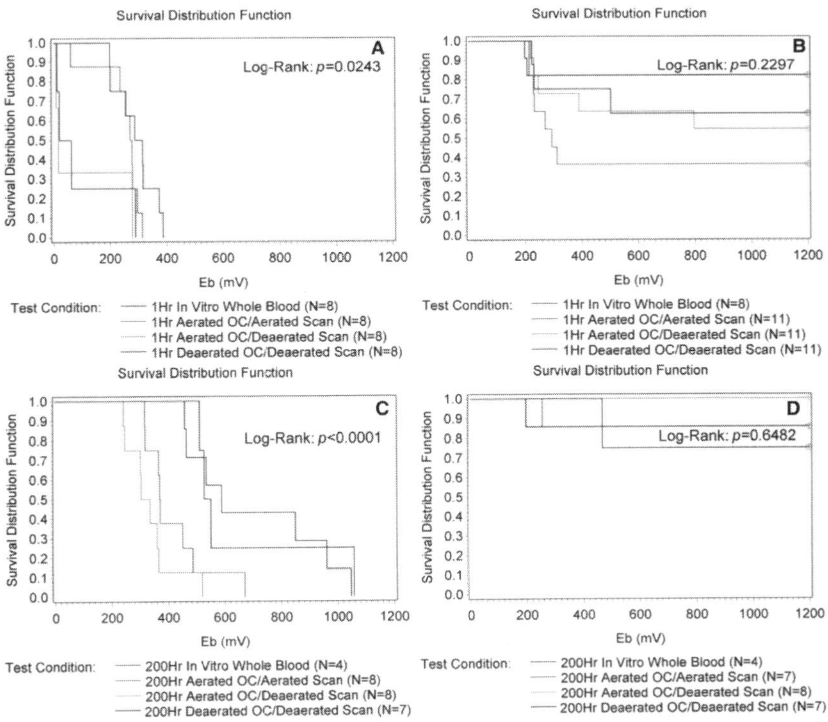

Figure 6. Kaplan–Meier curves for E_b: acute BO (**a**); acute EP (**b**); chronic BO (**c**); chronic EP (**d**)

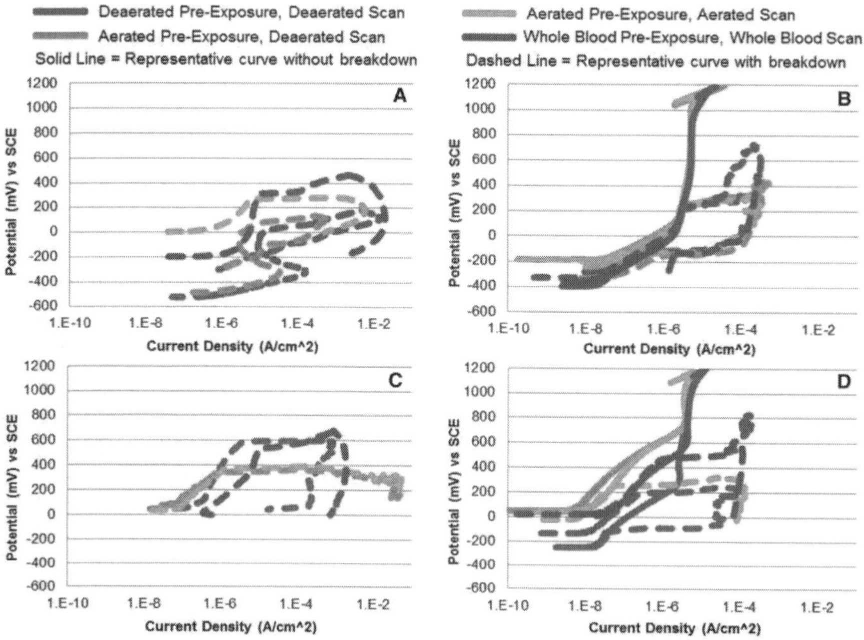

Figure 7. Representative cyclic polarization curves: acute BO (**a**); acute EP (**b**); chronic BO (**c**); chronic EP (**d**)

Therefore, this study investigated the effect of oxygen content and pre-exposure time on OCP, E_b, and E_b – OCP for two different NiTi oxide thicknesses (BO and EP). Both EP and BO samples were investigated since medical devices with EP and BO oxide thicknesses are currently available in the market. By analyzing two extremes in oxide thicknesses, a better understanding of how test conditions affect corrosion results for a broad range of NiTi medical devices can be achieved. Samples were tested in vivo (acute OCP only) and in vitro in whole blood to generate baseline physiologic results to compare against simulated fluid (aerated and deaerated PBS) results.

This study showed that both oxygen content and pre-exposure time had a significant effect on the corrosion behavior of NiTi, regardless of oxide thickness. Performing testing in vivo allows full understanding of the physiologic corrosion behavior. In this study, in vitro and in vivo whole blood resulted in similar OCP. Although not investigated in this study, it is expected that in vitro E_b and cyclic polarization curves in whole blood would be similar to that occurring in vivo. However, testing in whole blood alone is undesirable, due to the number of samples needed to establish reliability and the increased cost of testing. This study determined that 1 h data in whole blood were more conservative (i.e., lower E_b – OCP) than chronic data for both BO and EP. Therefore, a reasonable compromise might be to test a small sample size in whole blood for 1 h to gain an understanding of the representative physiologic corrosion behavior of the implant and compare those results to larger scale testing performed in a simulated solution.

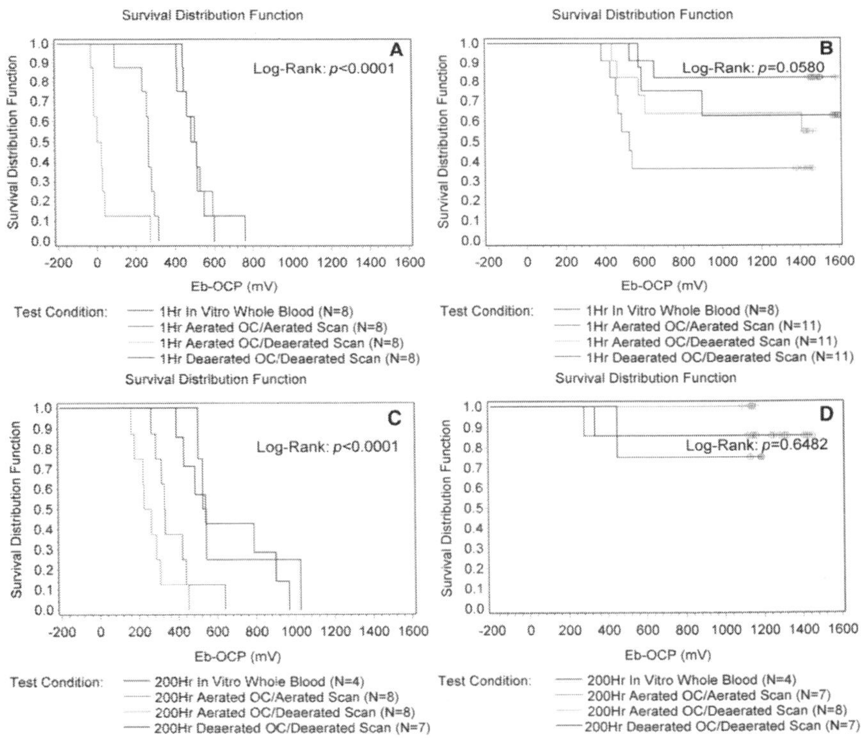

Figure 8. Kaplan–Meier curves for E_b – OCP: acute BO (**a**); acute EP (**b**); chronic BO (**c**); chronic EP (**d**)

Previous investigators have shown that E_b – OCP remained constant or increased with time, suggesting that 1 h pre-exposure time was conservative [8]. This is in agreement with the results obtained in this study. OCP, E_b, and E_b – OCP improved with time for both oxide thicknesses and aerations.

Table 3. Summary of results

Result	Figure
Open circuit potentials (OCP)	
In vitro whole blood OCP was not significantly different than in vivo OCP for both oxide thicknesses ($p = 0.0636$ and $p = 0.9964$)	Figure 2
Control (ASTM F2129) OCP was significantly different than in vivo OCP, depending on oxide thickness ($p < 0.0001$ for BO)	Figure 2
All OCPs increased over time	Figure 2
Chronic deaerated PBS samples were significantly different than chronic whole blood, depending on the oxide thickness ($p < 0.0001$ for EP)	Figure 2
Chronic-aerated PBS samples were not significantly different than chronic whole blood regardless of the oxide thickness ($p = 0.9976$ and $p = 0.7397$)	Figure 2
Breakdown potentials (E_b)	
All E_b improved over pre-exposure time	Figure 6
E_b tested per ASTM F2129 were significantly different than E_b chronically submerged in whole blood depending on oxide thickness ($p < 0.0001$ for BO)	Figures 6 and 7
Aerated $E_b \leq$ whole blood E_b for both oxide thicknesses and pre-exposure times, and had similar cyclic polarization curves	Figures 6 and 7
Difference in E_b – OCP	
All E_b – OCP values improved over pre-exposure time	Figure 8
Acute pre-exposure to aerated PBS prior to testing in deaerated PBS resulted in negative E_b – OCP values with no pitting observed during open circuit	Figures 8 and 9
Deaerated PBS and whole blood were comparable and the least conservative (i.e., the highest difference between E_b and OCP) test conditions regardless of pre-exposure time and oxide thickness. However acute cyclic polarization curves in deaerated PBS and whole blood were different, depending on oxide thickness	Figures 7 and 8
Aerated PBS was more conservative (i.e., the lower difference between E_b and OCP) than or comparable to whole blood regardless of pre-exposure time and oxide thickness	Figure 8

For deaerated PBS testing conditions, E_b – OCP was comparable to whole blood for both oxide thicknesses and pre-exposure times. However, the actual values of OCP, E_b, and the cyclic polarization curves were not always similar to those obtained in whole blood. In fact, the chronic EP OCP for deaerated PBS remained ~200 mV more active than whole blood, and the acute BO E_b for deaerated PBS was found to be ~250 mV more negative than whole blood. Additionally, the cyclic polarization curves for acute deaerated pre-exposure were very different than the curves tested in whole blood, depending on the oxide thickness. The acute deaerated BO samples exhibited a passivation potential where no such behavior was observed in whole blood. Therefore, testing in a deaerated environment may not fully capture the physiologic corrosion behavior of the sample in a physiologically relevant environment.

Aerated pre-exposure prior to performing the cyclic polarization scan in a deaerated environment has been recommended previously and resulted in an increased OCP and a more conservative E_b – OCP than testing with a deaerated pre-exposure [7]. Similarly, an increase in OCP and a conservative E_b – OCP with aerated pre-exposure followed by a deaerated scan were observed in this study. However, acute aerated pre-exposure was not sufficient to prevent the OCP from immediately becoming more active during the deaeration of the solution with nitrogen (Fig. 9a), resulting in a polarization scan starting at a more negative potential than the recorded aerated OCP. This resulted in an extremely conservative E_b – OCP value, in which some values were negative (though corrosion was not observed during the open circuit). Therefore, testing in a deaerated environment following 1 h of aerated pre-exposure produces potentially misleading results. It was found that 200 h of pre-exposure in an aerated solution was sufficient to reach a steady state such that the OCP was not influenced when nitrogen was introduced (Fig. 9b). Furthermore, chronic-aerated pre-exposure followed by a deaerated scan achieved a representative OCP and a conservative E_b, resulting in a conservative E_b – OCP compared to whole blood (~390 mV difference for BO). Therefore, a chronic pre-exposure in an aerated environment prior to a deaerated scan is a conservative testing option; however, it is essential that the OCP reaches a steady state.

Samples were also evaluated in an aerated environment following an aerated pre-exposure. For both oxide thicknesses, acute-aerated OCPs were most comparable to chronic physiologic OCP values. For both time points and oxide thicknesses, aerated cyclic polarization curves and E_b

were similar to whole blood. Additionally, aerated E_b – OCP was less than or equal to whole blood for both time points and oxide thicknesses. Overall, corrosion properties (OCP, E_b, and E_b – OCP) increased with time, indicating that acute aerated testing is conservative compared to chronic pre-exposure durations. However, increasing the pre-exposure duration might be a useful option for determining actual chronic corrosion properties. Additionally, testing in an aerated environment allows for a direct comparison to galvanic testing per ASTM F3044 [13]. Regardless of pre-exposure duration, testing in an aerated environment following an aerated pre-exposure is a viable and conservative option for determining the physiologic corrosion behavior of NiTi implants used for medical devices.

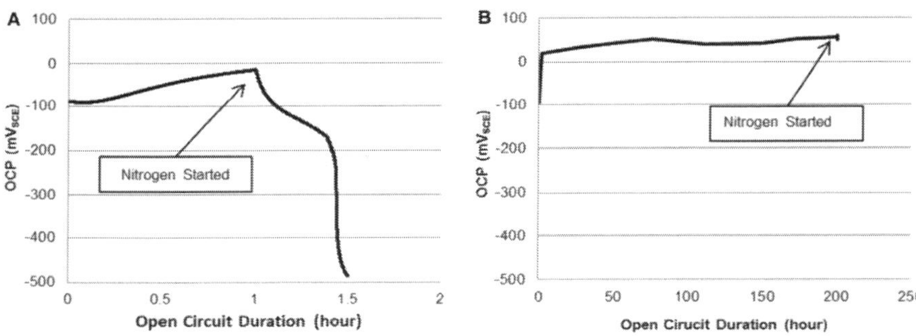

Figure 9. BO pre-exposure: 1 h aerated followed by 30 min deaeration with nitrogen (**a**), 200 h aerated followed by 30 min deaeration with nitrogen (**b**)

5. CONCLUSIONS

Whole blood can be used to perform in vitro testing to determine baseline corrosion behavior of medical implants and aid in acceptance criterion development. E_b – OCP tested according to the conditions defined in ASTM F2129 was comparable to acute whole blood and is conservative compared to chronic whole blood for both oxide thicknesses of NiTi. However, testing per the conditions in ASTM F2129 resulted in OCP, E_b, and cyclic polarization curves that were not always comparable to whole blood depending on oxide thickness. A standardized corrosion assessment method should not be dependent on metallic surface finish. Pre-exposing and conducting the polarization scan in aerated PBS achieved E_b, E_b – OCP, and cyclic polarization curves that were comparable to or more conservative than whole blood testing, regardless of pre-exposure duration and oxide thickness. Testing in aerated PBS also allows direct comparison to coupled galvanic potentials measured using the methods in ASTM F3044. Thus, testing in aerated PBS provides a more conservative and complete assessment of corrosion resistance of metallic medical implants.

ACKNOWLEDGMENTS

The authors would like to thank Sara Sherman, Jacqueline Garrison, and Scott Snyder of Cook Research Incorporated and William Voorhees, III of MED Institute, Inc. for their critical review of the manuscript. Additionally, the authors would like to thank Bret Teany and Alan Saunders of Cook Research Incorporated for their help in figure formatting. The authors would also like to thank the members of the Non-Clinical Testing teams at Cook Research Incorporated who supported the in vivo and in vitro testing. Finally, the authors would like to acknowledge Evans Analytical Group (www.eag.com) for conducting the Auger analysis.

REFERENCES

1. ASTM-F2129 (2008) Standard test method for conducting cyclic potentiodynamic polarization measurements to determine the corrosion susceptibility of small implant devices. ASTM International, West Conshohocken

2. Takai E (2012) Cardiovascular metallic implants: corrosion, surface characterization, and nickel leaching. FDA CDRH, p 36–78

3. Pértile LB, Silva P, Peccin VB, Peres R, Silveira PG, Giacomelli C, Giacomelli FC, Fredel MC, Spinelli A (2009) In vivo human electrochemical properties of a NiTi-based alloy (Nitinol) used for minimally invasive implants. J Biomed Mater Res Part A 89:1072–1078

4. Shih C-C, Lin S-J, Chung K-H, Chen Y-L, Su Y-Y (2000) Increased corrosion resistance of stent materials by converting current surface film of polycrystalline oxide into amorphous oxide. J Biomed Mater Res 52:323–332

5. Rosenbloom SN, Corbett R (2007) An assessment of ASTM F2129 electrochemical testing of small medical implants-lessons learned. Proceedings of the NACE Corrosion Conference & Exposition, Nashville, TN, USA, p 11–15

6. Corbett RA (2004) Laboratory corrosion testing of medical implants. Proceedings of Materials and Processes for Medical Devices Conference, p 166–171

7. Eiselstein L, Steffey D, Nissan A, Corlett N, Dugnani R, Kus E, Stewart S (2009) Acceptance criteria for corrosion resistance of medical devices: statistical analysis of nitinol pitting in in vivo environments. J Mater Eng Perform 18:768–780

8. Warner CP (2009) The effect of exposure to simulated body fluids on breakdown potentials. J Mater Eng Perform 18:754–759

9. Pound BG (2006) Susceptibility of nitinol to localized corrosion. J Biomed Mater Res Part A 77:185–191

10. Rondelli G, Torricelli P, Fini M, Rimondini L, Giardino R (2006) In vitro corrosion study by EIS of an equiatomic NiTi alloy and an implant quality AISI 316 stainless steel. J Biomed Mater Res B Appl Biomater 79:320–324

11. Istephanous N, Bai Z, Gilbert J, Rohly K, Belu A, Trausch I, Untereker D (2003) Oxide films on metallic biomaterials: myths, facts and opportunities. MSF 426:3157–3164

12. Guyton AC (2006) Textbook of medical physiology, Eleventh edn. Elsevier Health Sciences, Philadelphia

13. ASTM-F3044 (2014) Test method for standard test method for evaluating the potential for galvanic corrosion for medical implants. ASTM International, West Conshohocken

5

LONG-RANGE ATOMIC ORDER AND ENTROPY CHANGE AT THE MARTENSITIC TRANSFORMATION IN A NI-MN-IN-CO METAMAGNETIC SHAPE MEMORY ALLOY

Vicente Sánchez-Alarcos [1], Vicente Recarte [1,*], José Ignacio Pérez-Landazábal [1], Eduard Cesari [2] and José Alberto Rodríguez-Velamazán [3,4]

[1]Departamento de Física, Universidad Pública de Navarra, Campus de Arrosadía, 31006 Pamplona, Spain

[2]Departament de Física, Universitat de les Illes Balears, Ctra. de Valldemossa, km 7.5, E-07122 Palma de Mallorca, Spain

[3]Instituto de Ciencia de Materiales de Aragón, CSIC - Universidad de Zaragoza, 50009 Zaragoza Spain

[4]Institut Laue Langevin, F-38042 Grenoble, France

ABSTRACT

The influence of the atomic order on the martensitic transformation entropy change has been studied in a Ni-Mn-In-Co metamagnetic shape memory alloy through the evolution of the transformation temperatures under high-temperature quenching and post-quench annealing thermal treatments. It is confirmed that the entropy change evolves as a consequence of the variations on the degree of $L2_1$ atomic order brought by thermal treatments, though, contrary to what occurs in ternary Ni-Mn-In, post-quench aging appears to be the most effective way to modify the transformation entropy in Ni-Mn-In-Co. It is also shown that any entropy change value between around 40 and 5 J/kgK can be achieved in a controllable way for a single alloy under the appropriate aging treatment, thus bringing out the possibility of properly tune the magnetocaloric effect.

KEYWORDS

Ni-Mn-In-Co; entropy change; atomic order; martensitic transformation; metamagnetic shape memory alloys

1. INTRODUCTION

Ni-Mn-based Heusler alloys exhibiting both long-range magnetic ordering and thermoelastic martensitic transformation (MT) have been intensively investigated in recent years, from both fundamental and applied points of view, due to the unique properties they show linked to the occurrence of a first-order martensitic transformation (MT) between magnetically ordered phases [1]. The magnetism in these alloys mainly arises from the coupling between the Mn atoms (in which the magnetic moment is chiefly confined), so the magnetic exchange interactions strongly depend on the Mn-Mn distance [2], and therefore different sequences of magnetostructural transformations can be observed depending on both the X element and the change in interatomic distances caused by the MT. In particular, in the so-called metamagnetic shape memory alloys, $Ni_2Mn_{1-x}Z_{1+x}$ (Z=In, Sn, and Sb), the MT takes place between ferromagnetic austenite and a weak magnetic martensitic phase due to weakening of the exchange interactions (competing ferromagnetic and antiferromagnetic) [3–8]. The large change of magnetization, ΔM, occurring at the MT in this case allows the induction of the MT by an applied magnetic field, thus giving rise to multifunctional properties (namely giant magnetoresistance [9–12] magnetic shape memory [13], or large inverse magnetocaloric effect [14,15] of great technical interest for practical applications in sensing and magnetic refrigeration.

The magnetostructural MT in metamagnetic shape memory alloys is mainly driven by lattice dynamics and magnetism. Since the free energy of the ferromagnetic phase depends on the magnetization, a magnetic term appears in the energy balance between austenite and martensite due to the variation of the magnetic exchange coupling at the MT. In this sense, the entropy change linked to the MT can be considered as the sum of a vibrational term and a magnetic term ($\Delta S \approx \Delta S_{vib} + \Delta S_{mag}$) as long as there is no configurational contribution to the entropy change (due to the diffusionless character of the MT) and the electronic contribution is expected to be very small in these alloys [16]. The magnetic contribution, and therefore the entropy change at the MT, can be modified by the application of an external magnetic field [17] and, taking into account that the magnetic moment depends on the configurational ordering of the magnetic atoms in the crystal lattice, by the variation of the long-range atomic order. In this last respect, it has been recently shown that, for a given alloy composition, the atomic order can be easily modified in Ni-Mn-In alloys by means of different thermal treatments [18,19]. In particular, quenching from high temperatures (around the B2-$L2_1$ ordering temperature) allows to partially retain the low atomic order present at these temperatures, in such a way that the degree of long-range atomic order in the as-quenched alloys is lower than the equilibrium value allowed by the stoichiometry, which can be achieved after a slowly enough cooling. Likewise, the metastable state retained after quenching may evolve to the stable state (that is, the equilibrium atomic order degree of the alloy may be restored) once the alloy is heated up to temperatures at with atomic diffusion is possible, as it occurs in Ni-Mn-Ga alloys [20,21]. The increase in atomic order in Ni-Mn-In alloys enhances the magnetic character of the austenite, thus increasing its Curie temperature and lowering the MT temperature. Furthermore, the influence of atomic order on the transformation entropy points out the key role of atomic order on the thermodynamics of the MT in these alloys. In Ni-Mn-Sn and Ni-Mn-Sb, in turn, the $L2_1$ ordering is stable up to very high temperatures and no variation on the atomic order can be brought by thermal treatments [22].

The addition of Cobalt to Ni-Mn-In has been shown to enhance the magnetism of the austenite and to avoid ferromagnetic ordering in martensite [23,24]. This leads to an increase of ΔM at the MT which favors larger magnetically-induced shifts of the transformation temperatures and results in a higher associated magnetocaloric effect [25]. In these quaternary alloys, the entropy change at the transformation, ΔS, strongly varies with the degree of atomic order, which can be tailored by means of the thermal treatment [26]. The observed evolution of ΔS has been correlated to the difference between the Curie and martensitic transformation temperatures, which also depends on composition. However, up to now, no systematic study on the study of the effect of thermal treatments on the transformation entropy in Ni-Mn-In-Co alloys has been performed. In the present work, the influence of the atomic order on the martensitic transformation entropy change has been studied in a Ni-Mn-In-Co metamagnetic shape memory alloy through the evolution of the transformation temperatures under high-temperature quenching and post-quench annealing thermal treatments. It is confirmed that the entropy change evolves as a consequence of the variations on the degree of $L2_1$ atomic order brought by thermal treatments, though, contrary to what occurs in ternary Ni-Mn-In, post-quench aging appears as the most effective way to modify the transformation entropy in Ni-Mn-In-Co. It is also shown that any entropy change value between around 40 and 5 J/kgK can be achieved in a controllable way for a single alloy under the appropriate aging treatment, thus bringing out the possibility of properly tune the magnetocaloric effect.

2. EXPERIMENTAL SECTION

A polycrystalline $Ni_{45}Mn_{36.7}In_{13.3}Co_5$ (at %) ingot was prepared from high purity elements by arc melting under protective Ar atmosphere. The ingot was remelted several times and then homogenized in vacuum quartz ampoules at 1,173 K during 24 h. Small samples for calorimetric and magnetic measurements were obtained from discs previously cut from the center of the ingots by a slow speed diamond saw, and powder samples for neutron diffraction experiments were elaborated by crushing the samples in an agate mortar. Differential Scanning Calorimetry (DSC) measurements at a heating/cooling rate of 10 K/min were carried out in a TA Q100 DSC to study the thermal behavior of the samples. In order to analyze the effect of high-temperature thermal treatments on the transformation temperatures and the transformation entropy, the alloy was subjected to a 30 minutes annealing treatment at different temperatures (ranging from 773 K to 1173 K) followed by quenching in ice water. On the other hand, thermal cycles through the MT after heating up to different temperatures were performed on the as-quenched samples in order to observe "in situ" the evolution of the Curie and MT temperatures with the post-quench thermal treatments. The atomic order was analyzed by neutron diffraction measurements carried out at the Institute Laue-Langevin (Grenoble, France). In particular, powder neutron diffraction measurements were performed at a wavelength of 1.28 Å.

3. RESULTS AND DISCUSSION

Figure 1 shows the DSC thermograms obtained on a cooling-heating ramp in the vicinity of the martensitic transformation of the as-quenched samples. As indicated, the exothermic and endothermic peaks correspond to the forward and reverse martensitic transformations, respectively, and the λ-type shoulders are associated to the magnetic ordering of the austenite at the Curie temperature, T_C. In order to track the evolution of the MT, the temperatures of the MT peaks maxima, T_m^{for} and T_m^{rev}, have been taken as the forward and reverse MT temperature, respectively.

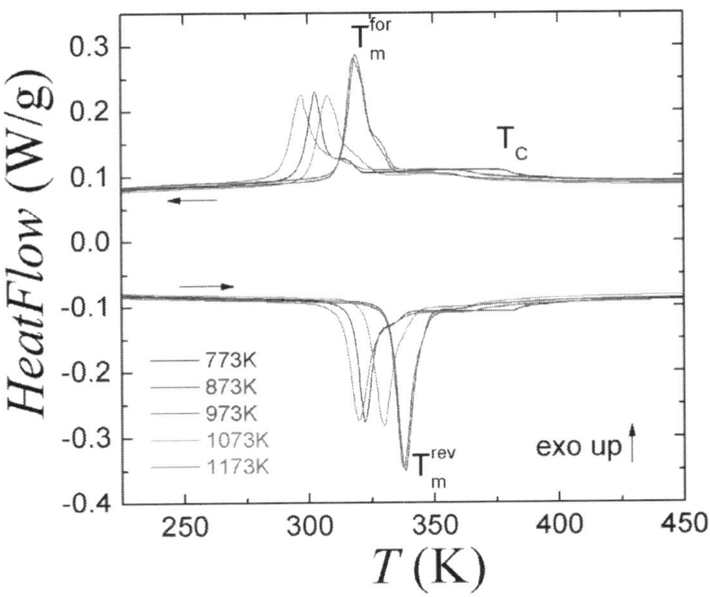

Figure 1. DSC thermograms corresponding to quenching from different temperatures.

As shown in Figure 2a, both the MT and Curie temperatures evolve with the quenching temperature, T_{quench}, though showing a non-monotonous behavior. In particular, the MT temperature T_m increases and T_C decreases with the increasingT_{quench} for T_{quench} < 900 K, and the opposite behavior, that is, a decrease on T_m and a slight increase on T_C, is observed when quenching from above 900K. The evolution of the transformation temperatures for T_{quench} < 900 K is similar to that found in ternary Ni-Mn-In, and can be ascribed to a decrease of the retained degree of $L2_1$ atomic order on increasing the quenching temperature [19]. On the other hand, assuming that quenching mainly affects the transformation temperatures due to the modification of atomic order, the behavior of T_m and T_C for T_{quench} > 900 K could be related to an increase of the retained degree of atomic order with the increasing T_{quench}, promoted by a high vacancies concentration at the quenching temperature that assist the ordering process, as in fact occurs in Ni-Mn-Ga and Ni-Fe-Ga alloys [20,27]. In any case, the net variations of the MT temperature ($\Delta T_m \approx 20$ K) are much lower than those found in Ni-Mn-In subjected to exactly the same quenching treatments ($\Delta T_m > 60$ K) [19], thus pointing out either a lower effect of quenching on the retained atomic order or a lower influence of atomic order on the MT. From the latent heat estimated from the area bellow the peaks in Figure 1, Q, the corresponding entropy change at the MT has been calculated as $\Delta S = Q/T_m$. The absolute values of the entropy change at the forward and reverse MT (ΔS_{for} and ΔS_{rev}, respectively) are plotted as a function of T_{quench} in Figure 2b. Although the absolutes values of ΔS_{for} and ΔS_{rev} are slightly different (as a result of the combination of the strong dependence of the magnetic contribution ΔS_{mag} on the temperature and the hysteresis of transformation) the evolution is almost the same in both cases; ΔS increases with T_{quench} for T_{quench} < 900 K and slightly decreases with T_{quench} for T_{quench} > 900 K. From the values in Figure 2a, it can be seen that the higher is the difference T_C-T_m the lower the entropy change, in agreement with previous results on similar alloys [26]. In this sense, the lower net variations of ΔS under quenching comparing with ternary Ni-Mn-In ($\Delta\ (\Delta S_{NiMnInCo}) \approx 10\ J/kgK$ vss $\Delta\ (\Delta S_{NiMnIn}) \approx 16\ J/kgK$) can be understood as a direct consequence of the smaller effect of quenching on the transformation temperatures.

In order to analyze the effect of thermal treatments on the atomic order, the temperature dependence of long range atomic order has been evaluated from neutron powder diffraction measurements. Figure 3a shows the neutron diffractogram obtained at 400 K (in paramagnetic austenite, to avoid magnetic contributions) on a powder sample quenched from 723 K. The alloy displays the cubic $L2_1$ structure typical of Heusler compounds, with lattice parameter a=5.9692 Å. The intensity peaks have been indexed according to the associated Bragg reflections.

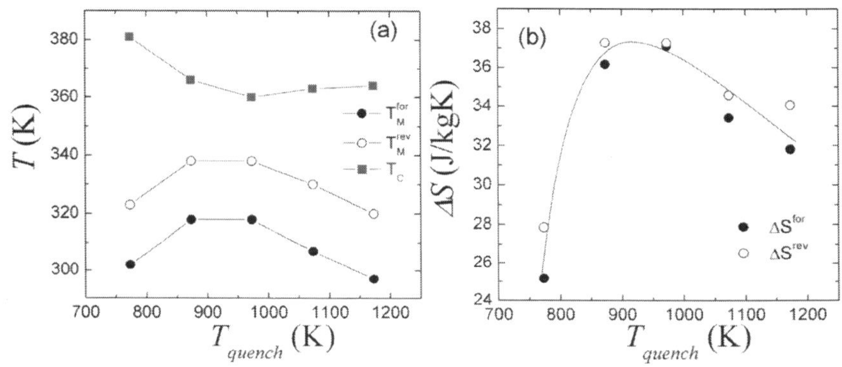

Figure 2. Transformation temperatures (**a**) and absolute value of the entropy change at the MT (**b**) as a function of the quenching temperature.

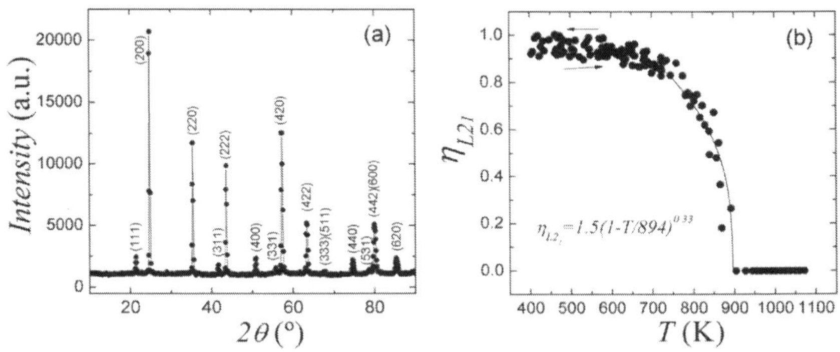

Figure 3. (**a**) Neutron diffractogram obtained at 400 K on a powder sample quenched from 723 K. (**b**) Temperature dependence of the $L2_1$ order parameter.

The evolution of the $L2_1$ atomic order with temperature has been estimated from the evolution of the integrated intensity of the austenitic (111) reflections, which are exclusively linked to that type of ordering (according to the corresponding unit cell structure factor, the $L2_1$ superstructure reflections are those in which the h, k, l indices are all odd). In particular, the order parameter corresponding to the $L2_1$ structure, η_{L21} (which ranges from 0, for the B2 structure, to 1, for the maximum next-nearest-neighbor atomic order allowed by the stoichiometry), has been determined assuming $\eta_{L21} \sim I_{111}^{1/2}$ and normalizing to the I_{111} value measured at 400 K after slowly cooling from 1,170 K [28]. Figure 3b shows the temperature dependence of the $L2_1$ order parameter on heating-cooling between 400 K and 1,173 K. The degree of $L2_1$ atomic order progressively decreases on heating above 700 K, until it becomes zero (the intensity of the (111) reflection completely vanishes) at around 900 K, thus evidencing the occurrence of a $L2_1 \leftrightarrow B2$ ordering transition. Since this transition has a second-order character, the behavior of η_{L21} when approaching the transition has been fitted to the power law $\eta_{L21} = A\ (1-T/T_{ord})^{\beta}$, obtaining values of $T_{ord} = 894\ K$, $A = 1.5$ and $\beta = 0.33$. The values of A and β parameters are consistent with the theoretical ones for a three-dimensional Ising model [29], and the value of the transition temperature T_{ord} is similar to that observed in the corresponding ternary $Ni_{50}Mn_{37}In_{13}$ alloy [30]. The degree of atomic order is close to unity below 750 K and becomes irrelevant above 900 K. Besides, a good agreement between heating and cooling curves is observed. Then, the quenching temperatures where substantial differences could be expected correspond to the 750–900 K temperature range. On the other side, the vacancies concentration at the different temperatures and their evolution must also play an important role during the quenching process and on the final order degree obtained at room temperature. So, the low influence of quenching on the transformation temperatures and on the MT entropy change can be explained since the atomic disorder retained by quenching is almost the same in all samples (except at 753 K where a large order degree is obtained).

Nevertheless, the atomic order can be better correlated with T_m, T_C and ΔS by means of post-quenched thermal treatments, taking into account that the equilibrium atomic order degree of the alloy may be restored once the alloy is heated up to a temperatures where atomic diffusion is possible. As previously reported in Ni-Mn-Ga and Ni-Mn-In alloys, the metastable-to-stable post-quench ordering process gives rise to an exothermic peak on the first heating curve of the DSC thermograms which makes it possible to identify and analyze the ordering process from calorimetric measurements. The exothermic peak associated with the post-quench ordering process taking place in the sample quenched from 1,073 K can be seen in the thermogram shown in Figure 4a. The inset shows a detail of the exothermic peaks appearing for different quenching temperatures. In order to determine the effect of the post-quench ordering process on the transformation temperatures and the entropy change, several consecutive heating/cooling DSC thermal cycles have been performed on the as-quenched samples. The cycles have been carried out at 10 K/min from a temperature below the MT to a temperature higher in each new cycle, in such a way that each cycle can be considered as a new aging treatment. As shown in Figure 4b, this procedure makes it possible to observe "in situ" the evolution of T_m and T_C as a consequence of the partial development of the ordering process.

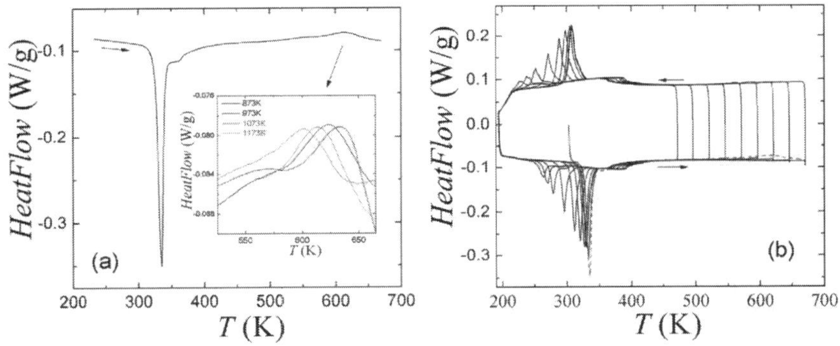

Figure 4. (**a**) DSC thermograms obtained on heating the sample just quenched from 1,073 K. Inset: detail of the exothermic peak associated to the ordering process in the samples quenched from different temperatures. (**b**) DSC thermograms performed on several consecutive thermal cycles on the sample just quenched from 1,073 K.

Figure 5a shows the increment on T_m^{rev} and T_C as a function of post-quench aging temperature, T_{aging} (that is, the maximum temperature of the DSC partial cycles). The DSC exothermic peak associated with the ordering process is also plotted in the figure. T_m^{rev} highly decreases ($\Delta T_m^{rev} \approx 70$ *K*) whereas T_C increases ($\Delta T_C \approx 30$ *K*) with the increasing T_{aging}, concurrently with the occurrence of the exothermic peak. This behavior is very similar to that observed in ternary Ni-Mn-In subjected to exactly the same thermal treatments, though the shifts are considerably lower in this latter case, $\Delta T_m^{rev} \approx 30$ *K*and $\Delta T_C \approx 15$ *K*, and it is in agreement with previous results on the effect of atomic order on the MT temperature in Co-doped Ni-Mn-In alloys [26]. The increase of the austenite Curie temperature as a result of the atomic ordering process is common to Ni-Mn-Ga and Ni-Mn-In alloys, and it origin relies on the enhancement of the ferromagnetic coupling between Mn atoms as the structure is getting ordered. Likewise, as it has been recently proposed, the enhancement of ferromagnetic coupling brought by the atomic ordering stabilizes the structural phase showing higher magnetic moment (in this case, the ferromagnetic austenite against the paramagnetic martensite) thus giving rise to a decrease of the MT temperature [31]. In this sense, the higher influence of post-quench aging on T_m and T_C observed in Ni-Mn-In-Co (with respect to the ternary alloy [19]) may be a consequence of both a higher retained atomic disorder, which would be in agreement with its lowerT_{ord}, and the higher saturation magnetization of the quaternary austenite. In any case, it is worth noting that the post-quench atomic ordering process results in a larger increase of the T_C-T_m difference, which leads to a considerably higher decrease of the entropy change at the MT ($\Delta(\Delta S_{NiMnInCo}) \approx 25$ *J/kgK vs* $\Delta(\Delta S_{NiMnIn}) \approx 7$ *J/kgK*), as shown in Figure 5b, where the increment of the entropy change at the MT with respect to the as-quenched state, $\Delta(\Delta S)$, is shown as a function of the post-quench aging temperature (for the sake of clarity, just the reverse MT is shown). With respect to the effect of atomic ordering on the transformation entropy ΔS, two main contributions, $\Delta S \approx \Delta S_{vib} + \Delta S_{mag}$, where ΔS_{vib} is the vibrational contribution and ΔS_{mag} is the magnetic contribution, can be considered (as already said, the electronic contribution ΔS_{el} is expected to be very small, even more if the alloy composition is not modified). In the case of the reverse MT, both terms in the entropy balance have opposite signs; $\Delta S_{vib} > 0$ due to the excess of vibrational entropy of the austenite with respect to the martensite [32], and $\Delta S_{mag} < 0$ as result of a MT from a magnetically disordered phase to an ordered one. Since the austenitic and martensitic overall structures do not evolve with atomic ordering, a similar variation on the density of vibrational states is expected in both phases, and consequently the effect of atomic ordering on ΔS_{vib} can be neglected in a first approximation, so the influence of atomic order on ΔS must be linked just to the change of ΔS_{mag} as a consequence of the variations in the magnetic exchange coupling in the austenitic phase.

The comparison between Figure 2b and Figure 5b confirms that post-quench aging is more effective than high temperature quenching to control the entropy change through atomic order variations, contrary to what occurs in the corresponding ternary Ni-Mn-In [18,19]. This is quite interesting since it means that almost any ΔS value between around *40 J/kgK* and *5 J/kgK* can be achieved in a controllable way for a single alloy. In fact, taking into account that the limit value of the magnetocaloric effect is the transformation entropy (if the effect of the magnetic field is strong enough to totally induce the first order transition) [33], this brings out the possibility of properly tune the magnetocaloric effect, which can be tailored in a continuous way by choosing the appropriate aging temperature. Furthermore, the range of achievable entropy values can be broadened by quenching the alloy from different temperatures. In this respect, Figure 6 shows the increments on T_m andT_C (Figure 6a) and the corresponding absolute values of the entropy change achievable at forward and reverse MT (Figure 6b) for the alloy quenched from 873 K, 973 K, 1,073 K and 1,173 K (from which a similar degree of atomic order is retained) and followed by cyclic post-quench ageing. It can be seen that, in fact, a very large range of entropy change values are obtained for a single alloy from the combination of quenching and post-quench treatments. As stated above, the evolution of the transformation entropy under thermal treatments is caused by the evolution of the magnetic contribution ΔS_{mag} as a consequence of the effect of long-range atomic order on the alloy magnetism.

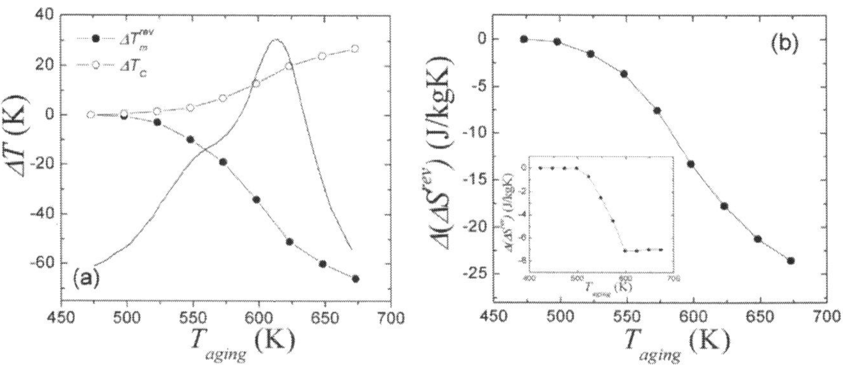

Figure 5. Increment on $T_m{}^{rev}$ and T_C (**a**) and increment on the transformation entropy at the reverse MT (**b**) in the sample quenched from 1,070K, as a function of post-quench aging temperature. Inset: Increment on the transformation entropy in Ni-Mn-In [19].

The role of the magnetic contribution to the transformation entropy becomes evident from the common dependence of the measured ΔS values on the normalized temperature difference, $\delta_A = (T_C - T_m)/T_C$, shown in Figure 7. This behavior agrees with the phenomenological model based on a Bragg-Williams approximation recently proposed to describe the evolution of the magnetic contribution to the total entropy change in Ni-Mn-based alloys [34] (a similar behavior has been also reported in Ni-Fe-Ga ferromagnetic shape memory alloys [35]). The absence of a marked change in the slope (previously reported) seems to be due to the paramagnetic nature of martensite in the present Ni-Mn-In-Co alloy. Furthermore, the huge set of values presented in the figure perfectly illustrates the above-mentioned ability of thermal treatments to modify the entropy change at MT as a result of the atomic order variations.

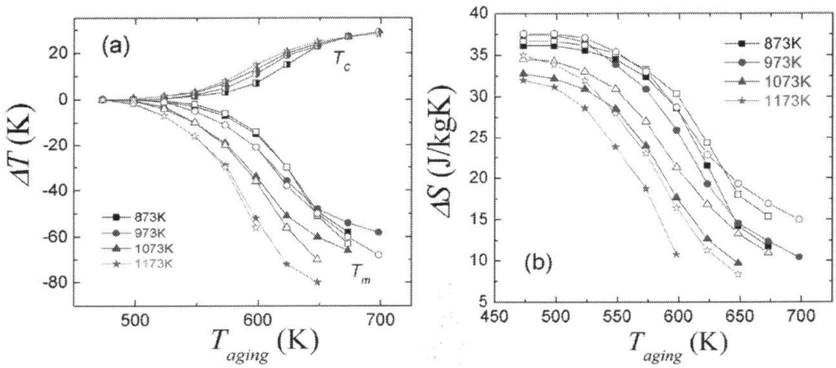

Figure 6. Increment on T_m and T_C (**a**), and increment on the absolute value of the transformation entropy (**b**), in the samples quenched from different temperatures and followed by cyclic post-quench ageing (for Tm and ΔS. closed symbols: direct MT; open symbols: reverse MT).

4. CONCLUSIONS

The influence of the atomic order on the martensitic transformation entropy change has been studied in a Ni-Mn-In-Co metamagnetic shape memory alloy. The entropy change evolves as a consequence of the variations on the degree of $L2_1$atomic order brought by thermal treatments. In particular, the evolution of ΔS is explained in terms of the variation of the magnetic contribution to the entropy as a consequence of the change in the magnetic exchange coupling in the austenitic phase. Contrary to what occurs in ternary Ni-Mn-In, post-quench aging appears as the most effective way to modify the transformation entropy in Ni-

Mn-In-Co. Almost any entropy change value between around 40 and 5 J/kgK can be achieved in a controllable way for a single alloy under the appropriate aging treatment, thus bringing out the possibility of properly tune the magnetocaloric effect.

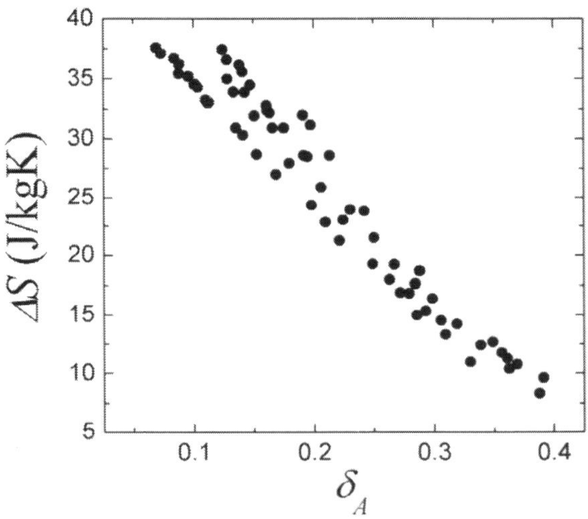

Figure 7. Dependence of the measured entropy change values on the normalized temperature difference, $\delta_A = (T_C - T_m)/T_C$.

ACKNOWLEDGMENTS

This work has been carried out with the financial support of the Spanish "Ministerio de Economía y Competitividad" (projects number MAT2011-28217 C02-01 and MAT2012-37923 C02-01). The Institute Laue Langevin-Spanish CRG D1B installation is acknowledged for the allocated neutron beamtime (Exp. CRG 1905).

AUTHOR CONTRIBUTIONS

Vicente Sánchez-Alarcos, Vicente Recarte, José Ignacio Pérez-Landazábal and Eduard Cesari wrote the manuscript. Eduard Cesari synthesized the samples. José Alberto Rodríguez-Velamazán performed temperature dependent neutrons diffraction experiments. Vicente Sánchez-Alarcos, Vicente Recarte, José Ignacio Pérez-Landazábal performed calorimetric measurements. All authors have read and approved the final manuscript.

REFERENCES

1. Kainuma, R.; Ito, W.; Umetsu, R.Y.; Khovaylo, V.V.; Kanomata, T. Metamagnetic shape memory effect and magnetic properties of Ni-Mn based Heusler alloys. *Mater. Sci. Forum* **2011**, *684*, 139–150.

2. Sasioglu, E.; Sandratskii, L.M.; Bruno, P. First-principles calculation of the intersublattice exchange interactions and Curie temperatures of the full Heusler alloys Ni_2MnX (X = Ga,In,Sn,Sb). *Phys. Rev. B* **2004**, *70*, 024427.

3. Sutou, Y.; Imano, Y.; Koeda, N.; Omori, T.; Kainuma, R.; Ishida, K.; Oikawa, K. Magnetic and martensitic transformations of NiMnX (X = In,Sn,Sb) ferromagnetic shape memory alloy. *Appl. Phys. Lett* **2004**, *85*, 4358–4360.

4. Krenke, T.; Acet, M.; Wassermann, E.F.; Moya, X.; Mañosa, L.; Planes, A. Martensitic transitions and the nature of ferromagnetism in the austenitic and martensitic states of Ni-Mn-Sn alloys. *Phys. Rev. B* **2005**, *72*, 014412.

5. Oikawa, K.; Ito, W.; Imano, Y.; Sutou, Y.; Kainuma, R.; Ishida, K.; Okamoto, S.; Kitakami, O.; Kanomata, T. Effect of magnetic field on martensitic transition of $Ni_{46}Mn_{41}In_{13}$ Heusler alloy. *Appl. Phys. Lett* **2006**, *88*, 122507.

6. Krenke, T.; Acet, M.; Wassermann, E.F.; Moya, X.; Mañosa, L.; Planes, A. Ferromagnetism in the austenitic and martensitic states of Ni-Mn-In alloys. *Phys. Rev. B* **2006**, *73*, 174413.

7. Khan, M.; Dubenko, I.; Stadler, S.; Ali, N. Exchange bias behavior in Ni-Mn-Sb Heusler alloys. *Appl. Phys. Lett* **2007**, *91*, 072510.

8. Aksoy, S.; Acet, M.; Deen, P.P.; Mañosa, L.; Planes, A. Magnetic correlations in martensitic Ni-Mn-based Heusler shape-memory alloys: Neutron polarization analysis. *Phys. Rev. B* **2009**, *79*, 212401.

9. Yu, S.Y.; Liu, Z.H.; Liu, G.D.; Chen, J.L.; Cao, Z.X.; Gu, G.H.; Zhang, B.; Zhang, X.X. Large magnetoresistance in single-crystalline $Ni_{50}Mn_{50-x}In_x$ (x = 14–16) upon martensitic transformation. *Appl. Phys. Lett* **2006**, *89*, 162503.

10. Sharma, V.K.; Chattopadhyay, M.K.; Shaeb, K.H.B.; Chouhan, A.; Roy, S.B. Large magnetoresistance in $Ni_{50}Mn_{34}In_{16}$alloy. *Appl. Phys. Lett* **2006**, *89*, 222509.

11. Wang, B.M.; Liu, Y.; Wang, L.; Huang, S.L.; Zhao, Y.; Yang, Y.; Zhang, H. Exchange bias and its training effect in the martensitic state of bulk polycrystalline $Ni_{49.5}Mn_{34.5}In_{16}$. *J. Appl. Phys* **2008**, *104*, 043916.

12. Pathak, A.K.; Gautam, B.R.; Dubenko, I.; Stadler, S.; Ali, N. Phase transitions and magnetoresistance in $Ni_{50}Mn_{50-x}In_x$Heusler alloys. *J. Appl. Phys* **2008**, *103*.

13. Du, J.; Zheng, Q.; Ren, W.J.; Feng, W.J.; Liu, X.J.; Zhang, Z.D. Magnetocaloric effect and magnetic-field-induced shape recovery effect at room temperature in ferromagnetic Heusler alloy Ni-Mn-Sb. *J. Phys. Appl. Phys* **2007**, *40*, 5523.

14. Krenke, T.; Duman, E.; Acet, M.; Wassermann, E.F.; Moya, X.; Mañosa, L.; Planes, A. Inverse magnetocaloric effect in ferromagnetic Ni-Mn-Sn alloys. *Nat. Mater* **2005**, *4*, 450–454.

15. Planes, A.; Mañosa, L.; Acet, M. Magnetocaloric effect and its relation to shape-memory properties in ferromagnetic Heusler alloys. *J. Phys. Condens. Matter* **2009**, *21*, 233201.

16. Khovaylo, V.V.; Oikawa, K.; Abe, T.; Tagaki, T. Entropy change at the martensitic transformation in ferromagnetic shape memory alloys Ni2+xMn1-xGa. *J. Appl. Phys* **2003**, *93*, 8483–8485.

17. Barandiarán, J.M.; Chernenko, V.A.; Cesari, E.; Salas, D.; Gutiérrez, J.; Lázpita, P. Magnetic field and atomic order effect on the martensitic

transformation of a metamagnetic alloy. *J. Phys. Condens. Matter* **2013**, *25*, 484005.

18. Recarte, V.; Pérez-Landazábal, J.I.; Sánchez-Alarcos, V. Dependence of the relative stability between austenite and martensite phases on the atomic order in a Ni-Mn-In Metamagnetic Shape Memory Alloy. *J. Alloy. Comp* **2012**, *536*, S308–S311.

19. Recarte, V.; Pérez-Landazábal, J.I.; Sánchez-Alarcos, V.; Rodríguez-Velamazán, J.A. Dependence of the martensitic transformation and magnetic transition on the atomic order in Ni-Mn-In Metamagnetic Shape Memory Alloys. *Acta Mater* **2012**, *60*, 1937–1945.

20. Sánchez-Alarcos, V.; Pérez-Landazábal, J.I.; Recarte, V.; Cuello, G.J. Correlation between atomic order and the characteristics of the structural and magnetic transformations in Ni-Mn-Ga shape memory alloys. *Acta Mater* **2007**, *55*, 3883–3889.

21. Sánchez-Alarcos, V.; Pérez-Landazábal, J.I.; Recarte, V.; Rodríguez-Velamazán, J.A.; Chernenko, V.A. Effect of atomic order on the martensitic and magnetic transformations in Ni-Mn-Ga ferromagnetic shape memory alloys. *J. Phys. Condens. Matter* **2010**, *22*, 166001.

22. Sánchez-Alarcos, V.; Pérez-Landazábal, J.I.; Recarte, V.; Lucia, I.; Vélez, J.; Rodríguez-Velamazán, J.A. Effect of high temperature quenching on the magnetostructural transformations and the long-range atomic order of Ni-Mn-Sn and Ni-Mn-Sb metamagnetic shape memory alloys. *Acta Mater* **2013**, *61*, 4676–4682.

23. Ito, W.; Imano, Y.; Kainuma, R.; Sutou, Y.; Oikawa, K.; Ishida, K. Martensitic and magnetic transformation behaviors in Heusler-type NiMnIn and NiCoMnIn metamagnetic shape memory alloys. *Metall. Mater. Trans* **2007**, *38*, 759–766.

24. Pérez-Landazábal, J.I.; Recarte, V.; Sánchez-Alarcos, V.; Gómez-Polo, C.; Cesari, E. Magnetic properties of the martensitic phase in Ni-Mn-In-Co metamagnetic shape memory alloys. *Appl. Phys. Lett* **2013**, *102*, 101908.

25. Kainuma, R.; Imano, Y.; Ito, W.; Sutou, Y.; Morito, H.; Okamoto, S.; Kitakami, O.; Kanomata, T.; Ishida, K. Magnetic-field-induced shape recovery by reverse phase transformation. *Nature* **2006**, *439*, 957–960.

26. Kustov, S.; Corró, M.L.; Pons, J.; Cesari, E. Entropy change and effect of magnetic field on martensitic transformation in a metamagnetic Ni-Co-Mn-In shape memory alloy. *Appl. Phys. Lett 94*, 01.

27. Santamarta, R.; Cesari, E.; Font, J.; Muntasell, J.; Pons, J.; Dutkiewicz, J. Effect of atomic order on the martensitic transformation of Ni-Fe-Ga alloys. *Scripta Mater* **2006**, *54*, 1985–1989.

28. Planes, A.; Mañosa, Ll.; Vives, E.; Rodriguez-Carvajal, J.; Morin, M.; Guenin, G. Neutron diffraction study of long-range atomic order in Cu-Zn-Al shape memory alloys. *J. Phys. Condens. Matter* **1992**, *4*, 553.

29. Essam, J.W.; Fisher, M.E. Padé approximant studies of the lattice gas and ising ferromagnet below the critical point. *J. Chem. Phys* **1963**, *38*, 802.

30. Miyamoto, T.; Ito, W.; Umetsu, R.Y.; Kainuma, R.; Kanomata, T.; Ishida, K. Phase stability and magnetic properties of $Ni_{50}Mn5_{0-x}In_x$ Heusler-type alloys. *Scripta Mater* **2010**, *62*, 151–154.

31. Sánchez-Alarcos, V.; Recarte, V.; Pérez-Landazábal, J.I.; Gómez-Polo, C.; Rodríguez-Velamazán, J.A. Role of magnetism on the martensitic transformation in Ni-Mn-based magnetic shape memory alloys. *Acta Mater* **2012**, *60*, 459–468.

32. Planes, A.; Mañosa, L. Vibrational properties of shape-memory alloys. *Solid State Phys* **2001**, *55*, 159–267.

33. Recarte, V.; Pérez-Landazábal, J.I.; Kustov, S.; Cesari, E. Entropy change linked to the magnetic field induced martensitic transformation in a Ni-Mn-In-Co shape memory alloy. *J. Appl. Phys* **2010**, *107*, 053501

34. Recarte, V.; Pérez-Landazábal, J.I.; Sánchez-Alarcos, V.; Zablotskii, V.; Cesari, E.; Kustov, S. Entropy change linked to the martensitic transformation in metamagnetic shape memory alloys. *Acta Mater* **2012**, *60*, 3168–3175.

35. Oikawa, K.; Omori, T.; Sutou, Y.; Morito, H.; Kainuma, R.; Ishida, K. Phase equilibria and phase transition of the Ni-Fe-Ga ferromagnetic shape memory alloy system. *Metall. Mater. Trans* **2007**, *338*, 767–776

6

THERMOMECHANICAL TREATMENTS FOR NI-TI ALLOYS

F.M. Braz Fernandes[1], K.K. Mahesh[1] and Andersan dos Santos Paula[2]

[1] CENIMAT/I3N, Departamento de Ciências dos Materiais, FCT/UNL, 2829-516 Caparica, Portugal

[2] Post-graduated Program in Metallurgical Engineering, UFF - Universidade Federal Fluminense, Volta Redonda, Brazil

1. INTRODUCTION

Thermomechanical treatments for shape memory alloys (SMA) are found to be one of the more economical, simpler, and efficient methods adopted for manipulating the transformation properties. The stability of phase transformation has been found to depend upon the thermomechanical treatments, such as hot- or cold-working, heat-treatment and thermal cycling. It has perhaps more important and wide reaching ramifications than many of the other stages in the fabrication of components and structures.

During the stages of preparation of SMA, hot working is adopted as one of processes in the form of rolling or drawing to incorporate the shape memory effect (SME). Such alloys can be directly employed for the applications. However, most of the times, the ingots are finally cold worked in the form of rolling or drawing before delivering to the application purpose. This allows the application engineers to subject the alloys to appropriate thermal/mechanical treatment in order to obtain the SMA with desired phase transformation properties. Hence, a sequence of cold work followed by heat treatment is considered to be a productive method to tailor the SME and superelasticity (SE).

In order to emphasize the various methods of thermal, mechanical, and thermomechanical treatments, the Chapter is divided into the following Sections and Sub-sections.

a) Cold working

b) Cold working followed by heat treatments

c) Effect of cooling rate during heat treatments

d) Hot working

e) Thermal cycling

f) Severe plastic deformation

g) High-pressure torsion (HPT)

h) Equal channel angular pressing (ECAP)

i) Concluding remarks

2. COLD WORKING

Cold working can induce dislocations and vacancies in the Nickel-Titanium (Ni-Ti) alloys. It is suggested that the possible mechanisms for the martensite stabilization in the Equiatomic Ni-Ti alloys come from deformed structures and deformation induced dislocations/ vacancies. Thermomechanical treatments of Ni-Ti SMA are important for the optimization of the mechanical properties and phase transformation characteristics. An important characteristic in the Ni-Ti SMA is the stability on direct and reverse transformations, related to the sequence and transformation temperatures, and thermal hysteresis [1-5]. The transformation temperatures in Ni-Ti SMA have been shown to be related to the presence of lattice defects introduced by cold working [6, 7]. Wu *et al.*, (1996) showed that the defects induced during cold working have the effect of suppressing the martensitic transformation and promoting the R-phase transformation [8]. The residual internal stress induced by cold-working defects is considered to be responsible for the R-phase transformation [9]. The deformation mechanisms and morphologies in polycrystalline martensitic CuZnA1 alloy have been examined by Adachi and Perkins [10]. They observed that a variety of deformation morphologies, including variant-variant coalesce, stress-induced martensite to martensite transformation, injection of foreign variants to plate groups, and internal twinning and slip, are all exhibited simultaneously in moderately cold-worked specimens.

Ni-Ti alloys have a wider application in the form of wires. Therefore, an understanding of the wire drawing properties is important. Thin oxide film with a smooth surface on TiNi wires can be used as a lubricant during the drawing process. However, thick oxide films which have cracks and spalling on the surface can be detrimental to the drawing surface and depress the shape memory effect and pseudoelasticity of TiNi SMA. MoS_2 is an effective lubricant for wire drawing of TiNi SMA [8]. Also, cold rolling has been one of the widely adopted processing techniques in order to obtain Ni-Ti alloy in the sheet form. In a study of the cold-rolled equiatomic TiNi alloy, it was found that the same phenomena of martensite stabilization appear, as reported in Cu-based shape memory alloys [9, 10]. It is well known that the martensite in the $Ti_{50}Ni_{50}$ alloy has 24 variants [11]. The variants will accommodate each other under thermal or mechanical stress. It is reasonable to suggest that the stress exerted by cold rolling causes the variants to accommodate, i.e., the stress forces the preferred orientated variants to accommodate the deformation strain in the favorable stress direction. An intensive study of the microstructure of the deformed martensite shows, in addition to the deformed martensite plates, a large number of dislocations and vacancies can also be induced during the cold rolling. These deformation-induced dislocations and vacancies have an important effect on the martensitic stabilization [9].

In Fig. 1, DSC thermograms of the Ni-Ti alloy plate subjected to 40% cold working are shown. When the cold worked specimen is heated from RT up to around 300°C, no phase transformation is observed and on the further heating, a broad upward peak appears around 350°C corresponding to recrystalization process. However, on cooling to RT, a clear exothermic peak appears around 75°C and that is attributed to A→M phase transformation. While heating again from RT, the DSC thermogram shows an endothermic peak around 85 °C corresponding to M→A phase transformation and while cooling the reverse phase transformation, A→M, is observed.

In Fig. 2, 3-d representation of the XRD profiles obtained at different temperatures from RT to 400°C for the 40% cold worked Ni-Ti specimen is shown. XRD spectra obtained at RT show the peaks corresponding to B19' structure, which are broad and with low intensity. As the temperature is increased, the peak corresponding to B2 structure starts to emerge around 190 °C and on further heating, the intensity of the peak increases. Broad and low intensity peaks are due to the deformation induced dislocations and vacancies which suppresses the martensitic transformation [12-14].

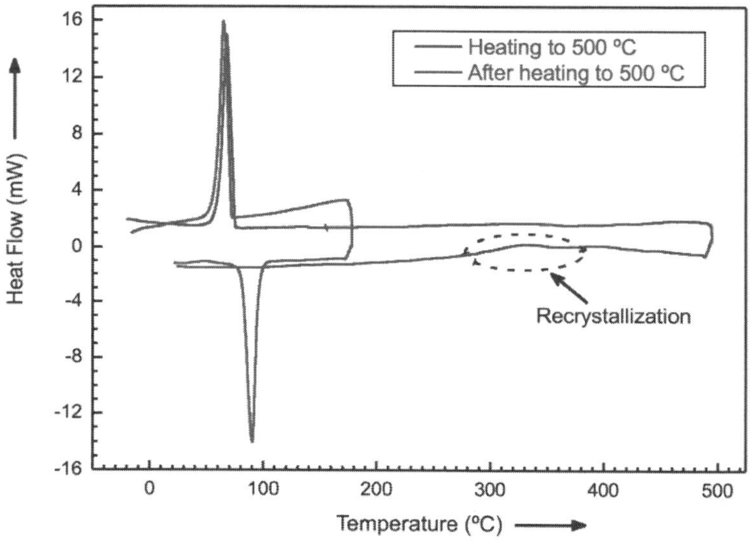

Figure 1. DSC thermograms of the Ni-Ti SMA plate initially cold worked up to 40% and heated to 500°C (in blue), and after (in red).

3. COLD WORKING FOLLOWED BY HEAT TREATMENTS

Heat treatment for metals and alloys has been proved to be an effective and economical process in order to maneuver their properties. Various factors, such as the HTT, annealing time and cooling rate after annealing have their own effects on the final state of the metal/alloy. In the above sub-section, it is mentioned that the defects induced during cold working have the effect of suppressing the martensitic transformation [8]. On the contrary, there is a possibility that a reverse phenomenon (restoration) would occur in a rather enhanced manner upon annealing through thermal activation processes of point defects. The migration of vacancies and interstitials could facilitate promotion of the martensitic transformation [15]. In this sub-section, the dependence of heat-treatment on the composition and thermal/mechanical history of the alloys has been explained. Heat-treatment plays a crucial role in fixing M_S. The detection of R-phase is found to be critical with the positioning of M_S in relation to R_S. If M_S is above R_S, R-phase is found to be masked by the martensite phase. Earlier, from electrical resistivity measurements, it was shown that while cooling from austenite phase, if R-phase exists, it preceded the martensite phase and it was regarded as the pre-martensitic phase [16]. However, later it was shown that both phases coexist at the same temperature, and it has been confirmed by the DSC study on the phase transformation in the 40% cold worked, near equi-atomic NiTi alloy subjected to water quenching from 400°C [17].

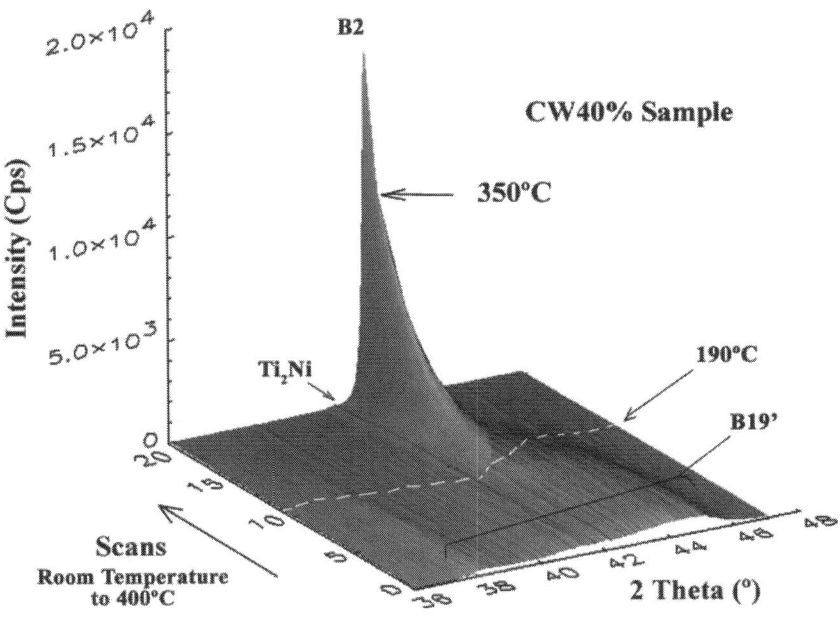

Figure 2. d representation of the XRD profiles obtained at different temperatures from RT to 400°C for the 40% cold worked Ni-Ti specimen.

Phase transformations associated with SME in Ni-Ti alloys can be one-stage, $B19^{I} \leftrightarrow B2$, two-stage including an intermediate R-phase stage, or multiple-stage depending on the thermal and/or mechanical history of the alloy. In a recent report, it has been highlighted the effect of (i) deformation by cold-rolling (from 10% to 40% thickness reduction) and (ii) final annealing on the transformation characteristics of a Ti-rich NiTi shape memory alloy. For this purpose, one set of samples initially heat treated at 500 °C followed by cold-rolling (10–40% thickness reduction) has been further heat treated at various temperatures between 400 and 800 °C. Phase transformations were studied using differential scanning calorimetry, electrical resistivity measurements and in situ X-ray diffraction. A specific pattern of transformation sequences is found as a result of combination of the competing effects due to mechanical-working and annealing [18].

Fig. 3 (a & b) show the Differential Scanning Calorimeter (DSC) and Electrical Resistivity (ER) curves for (i) as-received (AR), (ii) annealed at 500°C (HT500) and (iii) annealed at 500 °C/cold-rolled to 30%/annealed at 500°C (TMTCR30HT500) samples. For the AR sample, both in the case of DSC & ER techniques, multiple-step (B2↔R, B2↔B19', R↔B19', while heating and cooling) phase transformations are observed. For the HT500 sample, in both techniques, during heating and cooling, one-step (B19'↔B2) phase transformation is found to be present. Further, in the case of TMTCR30HT500 sample, one-step (B19'↔B2) phase transformation is detected. During heating (for AR samples), a small kink in the DSC and a small hump in ER plots around 60 °C show the presence of R-phase associated to multiple-step, (B19'↔R, B19'↔B2, R↔B2), phase transformation.

The effects of various heat treatment temperatures (HTT) on samples after being cold-rolled to different extents (10 to 40% thickness reduction) are presented in Fig. 4. All the samples were annealed at 500 °C before cold-rolling. Figs. 4 (a to d) show the transformation temperatures (A_f, A_s, R_{fh}, R_{sh}, R_{sc}, R_{fc}, M_s and M_f, obtained from DSC thermograms) as a function of HTT, for the samples annealed after being cold worked up to 10%, 20%, 30% and 40%, respectively. "A", "R" and "M" are the austenite, rhombohedral, and martensite phases; suffixes "s" and "f" are the start (1%) and finish (99%) transformation temperatures; and "c" and "h" refer to cooling and heating, respectively.

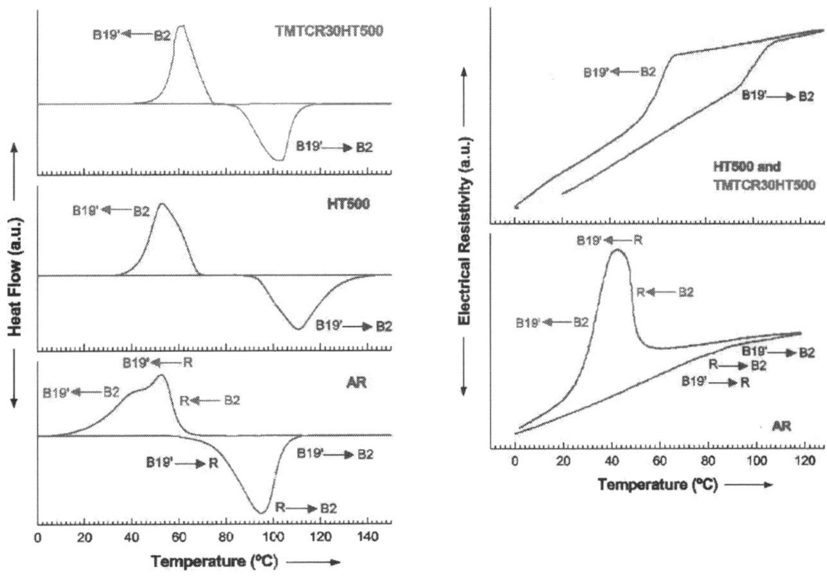

Figure 3. a) DSC and (b) ER curves for AR, HT500 and TMTCR30HT500 samples

In Fig. 4(a), for the 10% cold worked samples, as the final annealing temperature is increased, M_s and M_f are found to increase gradually up to 600 °C followed by a slight drop up to 800 °C. A_s and A_f are found to decrease as the final annealing temperature is increased from 400 to 500 °C. Further increase up to 700 °C shows gradual increase followed by a decrease for the final annealing temperature of 800 °C.

In Fig. 4(b), it is observed that for the 20% cold worked samples, there is R-phase formation while cooling (R_{sc}, R_{fc}) and while heating (R_{sh}). As the final annealing temperature is increased, R_{sc} and $R_{fc}$are found to increase till 500 °C. For higher final annealing temperatures, the R-phase formation is no longer detected. M_s and M_f increase with increasing final annealing temperature until it reaches 600 °C, followed by a slight decrease when the sample is heat treated at 700 °C. For the final annealing at 800 °C, M_s is not possible to be determined, but M_f increases slightly. A_s is found to increase with increasing final annealing temperature up to 700 °C along with A_f. For the final annealing temperature of 800 °C, A_s was not possible to be determined and A_f decreases. For this same treatment (800 °C), the R-phase formation is once again detected during cooling and heating.

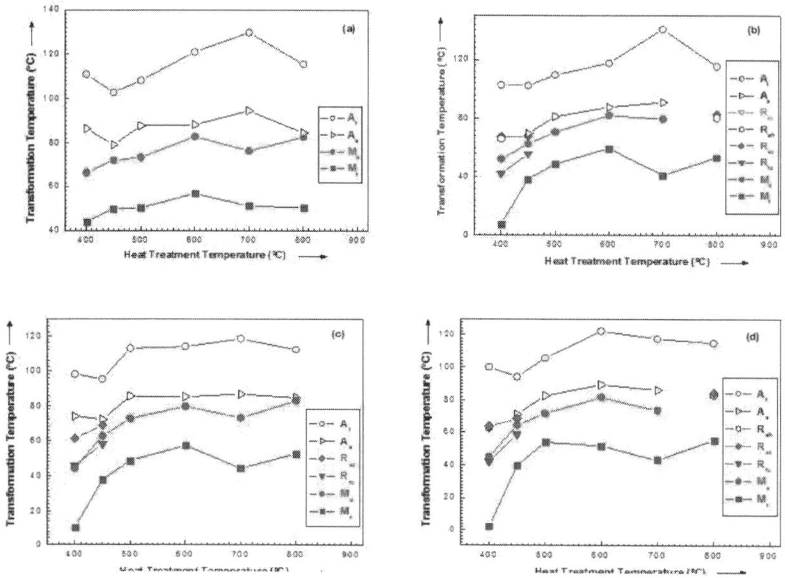

Figure 4. Transformation temperatures of (a) TMTCR10%, (b) TMTCR20%, (c) TMTCR30%, and (d) TMTCR40%

In the case of samples 30% cold worked, as shown in Fig. 4(c); the R-phase is only present during cooling for final annealing temperatures up to 500 °C (R_{sc} and R_{fc} increase with increasing final annealing temperature). M_s and M_f increase for increasing final annealing temperature up to 600 °C, slightly decrease for 700 °C and then slightly increase for 800 °C. A_s and A_f slightly decrease from 400 to 500 °C and then increase and stabilize after 500 °C.

In Fig. 4(d), it is observed that for the samples 40% cold worked and heat treatment there is R-phase formation only during cooling for the final annealing temperature up to 500 °C. (R_{sc} and R_{fc} are found to increase with increasing annealing temperature). M_s and M_f increase with increasing final annealing temperature till 600 °C. For the final annealing temperature of 800 °C, M_s and A_s were not possible to be determined. For the final annealing temperature of 800 °C, the R-phase formation is once again detected.

The absence of the R-phase formation in the sample annealed at 500 °C (not cold-rolled), may be explained by the annealing out of the structural defects and generation of the strain free crystals [19]. The same result is observed for the sample that has been cold-rolled to 10% (very close to the maximum recoverable strain of this class of alloys). With increasing extent of cold-work deformation, the R-phase deformation is only detectable for final annealing temperatures below 500 °C or at 800 °C. The final annealing temperature above 500 °C induces a recrystallization of the marformed matrix that makes the single-step transformation B2↔B19' more favorable [14, 20, 21]. This transformation may be initiated at the coherent interfaces of the very narrow precipitates Ti_2Ni. For the highest final annealing temperature (800 °C) the R-phase formation is once again present and this may be associated to the coalescence of the Ti_2Ni precipitates, making the B2 / Ti_2Ni interfaces incoherent [22, 23]. When the DSC and ER results in Figs. 1 and 2 are compared, it is apt to mention that when there is overlap of the phases transformation, ER technique is in a better position to reveal the presence of distinct phases.

Table 1 summarizes the transformation sequences of the samples after the thermomechanical treatments. For the samples cold worked to 10% and subsequently heat treated up to 700 °C, the transformation sequence is found to be clearly one-step (B19'↔B2). On the other hand, no matter the thickness reduction by cold-rolling, when the final annealing temperature is between 500 °C and 700 °C, the transformation is also clearly one-step (B19'↔B2). The two-steps phase transformation while cooling is only observed for the samples cold-rolled to 30 and 40% and for the final annealing temperatures of 400 °C. The multiple-steps phase transformation (with overlap) is only observed in two situations: (i) for the final annealing temperature of 800 °C, no matter the cold-work reduction, both while cooling and heating, and (ii) for the samples cold-worked to 20 to 40%, where the final annealing temperature was 500 °C or below.

Deformation up to 10% thickness reduction decreases the shape memory effect capability. This behavior is associated with the reorientation of martensite variants and increase of dislocation density, giving rise to a stabilization of martensite at a higher temperature in agreement with previous results [24].

Table 1. Influence of the thermomechanical processing (marforming) conditions on the transformations sequence.

HTT (°C)	Reduction by Cold Rolling			
	10%	20%	30%	40%
400	+/+	⊕/Ø	++/+	++/Ø
450	+/+	⊕/Ø	⊕/Ø	⊕/Ø
500	+/+	+/+	+/+	+/+
600	+/+	+/+	+/+	+/+
700	+/+	+/+	+/+	+/+
800	Ø/Ø	Ø/Ø	Ø/Ø	Ø/Ø

On Cooling / On Heating: + one-step; ++ two-steps;

⊕ Multiple-steps with overlap; Ø suspect multiple-steps with overlap.

4. EFFECT OF COOLING RATE DURING HEAT TREATMENTS

During the heat treatments, one of the parameters which, can be easily controlled is the cooling rate. Otsuka *et al.*, adopted a heat-treatment in which they homogenized the Ni50at%-Ti alloy for 1 h at 1000 °C followed by furnace cooling to eliminate the vacancies and the disorder to some extent. They found that quenched specimen has almost the same transformation temperatures as the furnace cooled one [25]. It was found earlier by Saburi *et al.*, that during heat-treatment, M_S and mechanical behavior of Ni-rich off-stoichiometric (>50.7at% Ni) NiTi alloys were sensitive to rate of cooling, whereas, of a near-stoichiometric (50.4 at% Ni) alloys were not [26]. Sitepu *et al.*, showed that precipitation of Ni_4Ti_3 particles occurred in a matrix of Ni-rich Ni-Ti SMA of nominal composition Ni50.7at%-Ti, when it was solution annealed at 850 °C for 15 minutes followed by water quenching and aging at 400 °C for 20 h [27]. In a more recent study, transformation behavior of NiTi alloys of different composition, heat treated by employing quenching and furnace cooling were investigated [28].

Fig. 5 shows resistivity profiles for the 2 samples, (a) Ni54.76wt%-Ti, i.e. Ti-rich and Ni56.00wt%-Ti, i.e. Ni-rich Ni-Ti alloys, in the as-received condition. For Ti-rich alloy, R-phase is found to occur only on cooling and the transformation is confined to a temperature interval of about 60º, above 0ºC. In the case of Ni-rich alloy, R-phase is found to appear both while heating and cooling, and its temperature interval is spread over a wide temperature range of more than 150º, below +50ºC, and these materials do not undergo the transformation to M-phase in the observed temperature range.

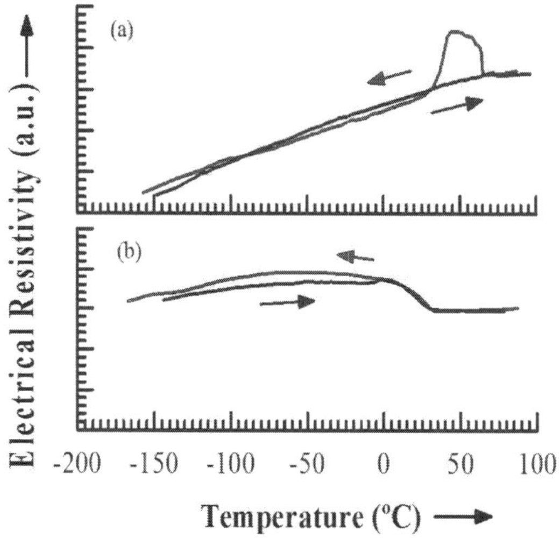

Figure 5. Electrical resistivity profiles for (a) Ni54.76wt%-Ti and (b) Ni56.00wt%-Ti alloys in the as-received condition

Fig. 6 (a-c) and 6 (d-f) show the resistivity profiles of the quenched and furnace cooled samples of Ti-rich alloy, respectively. In both cases, profiles are similar. R-phase transformation is only present during cooling for all the samples annealed between 100° and 420°C and the transformation region decreases, with increase in annealing temperature due to the increase in M_S temperature. For the annealing temperatures between 420°- 800°C, R-phase is found to be absent.

Fig. 7 (a-d) and 7 (e-h) demonstrate the resistivity profiles of the Ni-rich alloy for the quenched and the furnace cooled samples respectively. For the quenched samples, annealed in the temperature range of 100°-500°C, two-stage transformation A→R→M during cooling and M→R→A during heating are observed. When annealed between 500° and 600°C, two-stage transformation is observed only in cooling, with decrease in the temperature interval of R-phase. Annealing above 600°C, further suppression of R-phase takes place promoting only M↔A transformation. In the case of furnace cooled alloy, with increase in annealing temperature, a unique discontinuous behavior is observed. With increase of annealing temperature from 100º to 440ºC, two-stage transformation is observed both during cooling and heating in the resistivity profile, with reduced R-phase temperature interval. Annealing the sample between 440º and 580ºC, R-phase is found only on cooling with further reduction in the temperature interval. For the sample annealed at 590ºC, a sudden increase in the temperature interval of R-phase takes place. Hence, annealing around 590ºC seems to be very critical. Annealing above 590ºC, two-stage transformation is seen both during heating and cooling in the resistivity profile, regaining the initial behavior. The profiles indicate the stabilization of various phases above annealing temperatures of 590ºC.

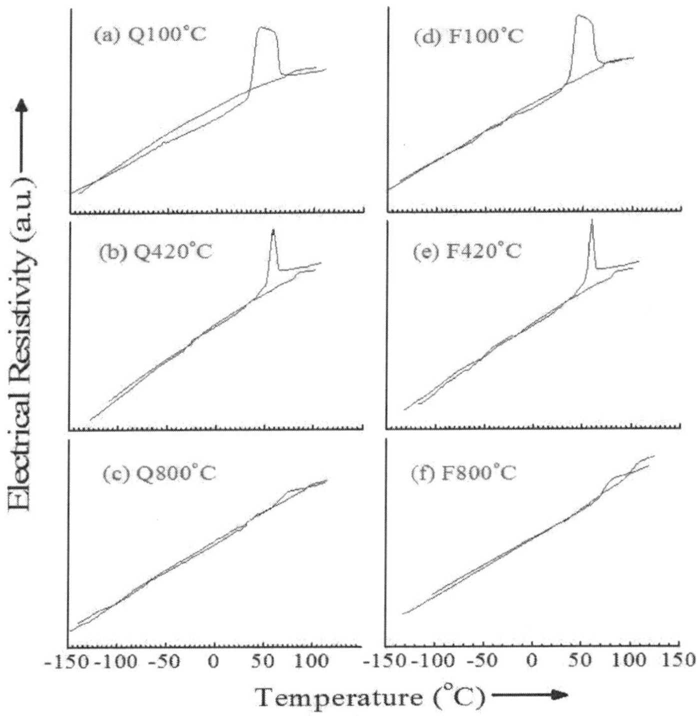

Figure 6. Resistivity profiles for the quenched and furnace cooled Ni54.76wt%-Ti alloys annealed at different temperatures.

For lower annealing temperatures, all the samples of the two alloys, both quenched and furnace cooled, exhibit similar behavior, i.e., M_s increases with increase in annealing temperature, which is attributed to the release of energy stored during the cold work. Cold work introduces high density of lattice defects, residual strain and internal stresses in the materials, which hinders from the movement of martensite interfaces. On annealing such cold worked materials, thermally activated diffusion leads to the annihilation of lattice defects, promoting martensitic transformation [29]. For the quenched samples, at higher annealing temperatures, this trend continues and gradual reduction in R-phase facilitates M↔A transformation. But, the furnace cooled samples, after annealing at higher temperatures, behave differently. A comparison of the resistivity profiles for the quenched and furnace cooled samples, especially annealed at higher temperatures, indicates that Ni-rich alloy is sensitive to the cooling procedure, unlike Ti-rich alloy. There is not much difference in the behavior of Ti-rich alloy either furnace cooled and quenched. In the case of furnace cooled Ni-rich alloy a unique discontinuous behavior is observed, for annealing at 590 °C. This may be due to the microstructural variations, arising as a consequence of two competing processes, viz., annihilation of defects and precipitation. Annealing above this critical temperature, the sample is able to regain and sustain a two-stage transformation, which may be attributed to the dominance of precipitation process over the defect annihilation process. It is proposed that, there is increased chance for Ti_3Ni_4 precipitation while furnace cooling, due to the slow cooling process and the presence of the material at higher temperature for a longer time. As reported by Nishida *et al.*, Ti_3Ni_4 precipitates have rhombohedral structure and are coherent to the matrix having a B2 type structure [30].

5. HOT WORKING

Both rolling temperature and thickness reduction are important factors that influence the work hardening and hardness of hot-rolled plates. The greater the thickness reduction, the greater the number of dislocations retained, and therefore, the greater the rate of work hardening. At rolling temperatures ≥600 °C, recovery or recrystallization occurs. However, because of the short rolling time and the fast cooling in air, the recovery or recrystallization is incomplete [31]. Hot-rolled Ni-Ti materials are found to possess enhanced resistance to low-cycle fatigue (increased pseudoelastic stability) as long as the primary material processing route remains unchanged [32]. Paula *et al.*, recently studied Ni-Ti alloys subjected to heat treatment at 767 °C for 300 s followed by hot rolling (50%) after cooling in air to 500 °C and water quenching to room temperature (T_{room}). Phase transformations were studied using differential scanning calorimetry, electrical resistivity measurements and in situ X-ray diffraction [18].

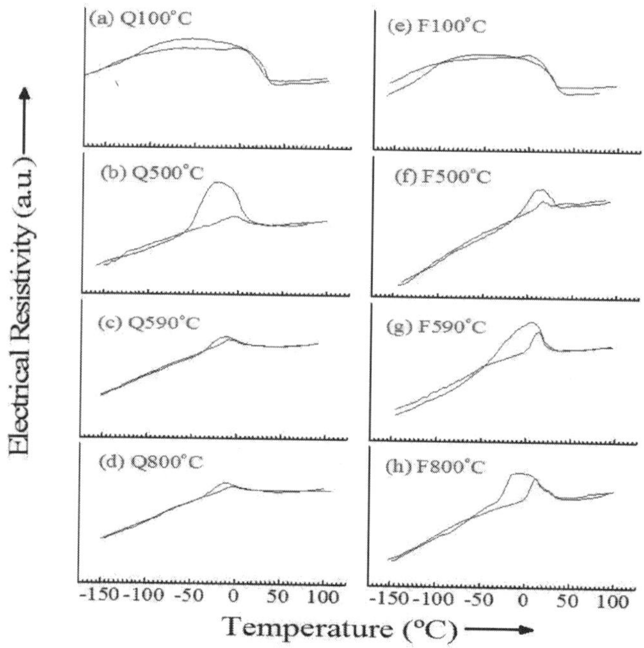

Figure 7. Resistivity profiles for the quenched and furnace cooled Ni56wt%-Ti alloys annealed at different temperatures.

Fig. 8 (a & b) shows the DSC and ER curves for the ausformed at 500 °C (TMTHR500) samples. During the cooling and heating stages, multiple-step (B2↔R, B2↔B19', R↔B19') phase transformation is clearly detected in both techniques. During heating, a small kink in the DSC and a small hump in ER plots around 60 °C show the presence of R-phase associated to multiple-step, (B19'→R, B19'→B2, R→B2), phase transformation. It was found that the ausforming at 500 °C promotes multiple-step phase transformation on cooling and heating (B2↔R; B2↔B19'; R↔B19'). During the ausforming process at 500 °C, it is not achieved a full recrystallization, in agreement with other authors results [33]. Ausforming introduce many defects in the sample, so that R-phase formation becomes necessary to decrease the energy for B2→B19' or B19'→B2 transformations.

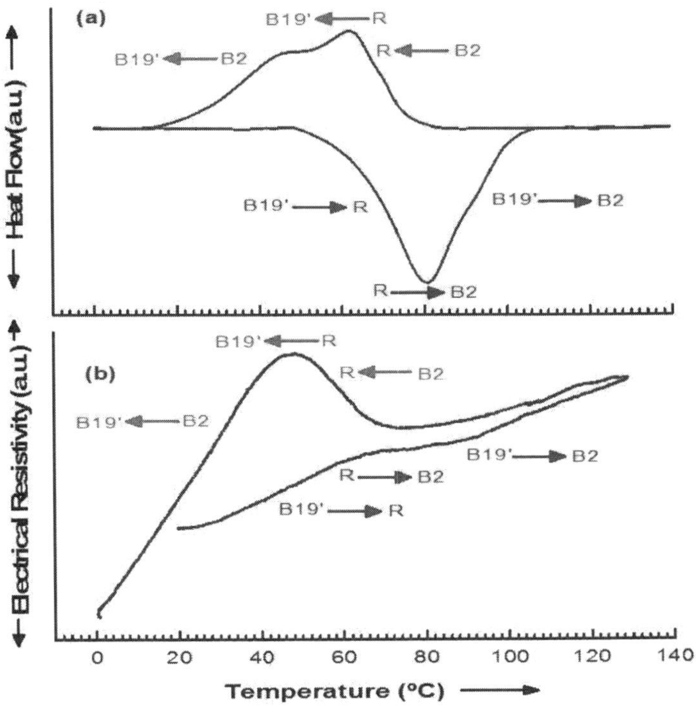

Figure 8. a) DSC and (b) ER curves for TMTHR500 samples.

6. THERMAL CYCLING

Thermoelastic martensitic transformation appears to be very sensitive to thermal cycling [34, 35]. Also, thermal and mechanical treatments can suppress slip deformation resulting in increase of flow stress and modify the transformation temperatures, recovery stresses and recovery strains [36]. These observations indicate that the transformation process is strongly affected by irreversible changes in the microscopic state of the alloy introduced by thermal cycling.

Thermal cycling causes a decrease in the characteristic temperatures and heats of transformation [37]. Also, thermal cycling is found to promote the intermediate R-phase transformation [38]. The effect of training conditions and extended thermal cycling on the two-way shape memory behavior of nitinol has been studied by **Hebda and White, 1995** [39]. Thermal cycling under constant load was studied by **de Araujo et al., 2000** [40] and they concluded that the internal stresses created were effective in inducing two-way memory effect.

Below, in Fig. 9, phase transformations are studied during the ab initio 10 thermal cycles by using DSC and ER techniques. In the DSC, thermal cycle was comprised of heating up to 140 °C, holding for 360 s and subsequently cooling down to -30 °C, with heating and cooling rates being 7.5 K/min. ER characterization have been performed by making use of a home made four-probe setup and the thermal cycling is performed by using the temperature controlled silicone oil bath. Ni-Ti (Ti51at%-Ni) alloy has been previously subjected to a series of thermomechanical treatment followed by heat treatment at 500 °C for 30 min. [41].

In Fig. 9 (a & b), during the first thermal cycle, in both the techniques (DSC & ER), it is observed that one-step phase transformation takes place. As the thermal cycling progresses, phase transformation processes are found to shift toward lower temperatures, both while heating and cooling. In Fig. 9(a),DSC thermograms for the first and second thermal cycles, the phase transformation peaks are observed to be symmetrical both while heating and cooling attributing to one-stage M↔A transformation. Also, in the ER profile shown in Fig. 9(b) corresponding to the first and second thermal cycles, it is observed that the specimen undergo one-step M↔A transformation. As the number of thermal cycles is increased, DSC thermogram peaks is found to broaden asymmetrically and shift toward lower temperatures (from the fifth cycle onward), giving rise to increasing evidence of the intermediate R-phase transformation while cooling (Fig. 9b).

This shows that the Ti-rich Ni-Ti alloy under study, when subjected to thermal cycling, after multiple steps of thermomechanical treatments followed by final heat treatments, the stability of the phase transformation is found to sensitive and depend on the final heat-treatment temperatures. Further, the thermal cycling process also found to affect the nature of phase transformation. Further, it can also be inferred that different thermomechanical treatments applied on a specimen are found to have opposing effects on the nature of phase transformations. In contrast to the heat treatments, which tend to increase the phase transformation temperatures, thermal cycling tends to decrease them.

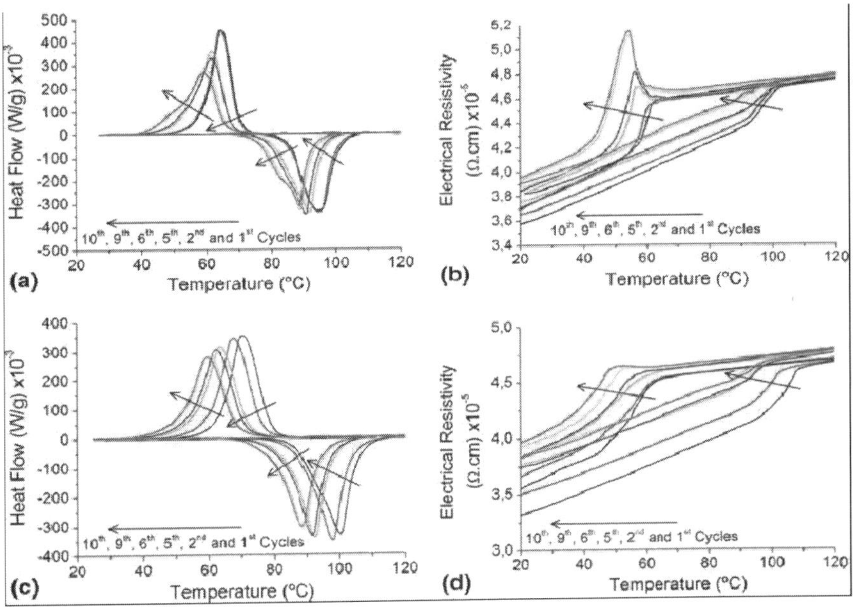

Figure 9. Evolution of phase transformation during thermal cycling up to 10 of the Ni-Ti specimens subjected to series of thermomechanical treatment followed by heat treatment at 500 °C for 30 min. (a) DSC and (b) ER profiles.

7. SEVERE PLASTIC DEFORMATION

The plastic deformations carried out by cold-working and hot-working presented above have been extended in the recent past, by subjecting these alloys to severe plastic deformation (SPD). It was shown that the effects of high density of grain boundaries on the martensitic phase transformation and the functional properties of SMA became a focus of research investigating the impact of ultrafine and nanograins on the parameters of the SME and SE. Further, methods of SPD, such as high pressure torsion (HPT) and equal channel angular pressing (ECAP) have been successfully applied to achieve ultrafine grained (UFG) and bulk nanostructured SMA [42–45].

a. High pressure torsion (HPT)

Waitz et al. [44] showed that martensitic transformation shifts to low temperature when the grain size is less than 150 nm. Initially in their experiments, Ni–Ti alloy was subjected to HPT and later annealed close to recrystallization temperature. By post-deformation annealing at 300°C, it was found that the amorphous structure created by the room-temperature HPT loses its thermomechanical stability and intensively crystallizes [45]. The effect of the composition on the phase transformations in Ni–Ti alloys subjected to HPT and followed by heat treatments was recently reported [46].

Bulk Ni-Ti SMA with different compositions have been chosen and subjected to HPT and their phase transformation characterization was carried out. The selected Ni(49.6 to 49.4at%)-Ti (Ti-rich) alloy in the as-received (AR) condition has M_f above RT and Ni(around 50.8at%)-Ti (Ni-rich) has A_f below RT. SPD of Ni-Ti alloys (Ti-rich and Ni-rich) have been performed by HPT at RT. Further, HPT processed separate specimens are subjected to heat treatments at temperatures of 300°C (HPT+HTT300) and 350 °C (HPT+HTT350) for 20 min, and quenched into water at room temperature. Phase transformation temperatures are analyzed by studying the Differential Scanning Calorimeter (DSC) plots. Further, the structural evolution of the samples subjected to SPD in the phase transformation temperature region was studied using in situ X-ray diffraction (XRD) from −180 to +180°C.

The phase transformation temperatures obtained from the thermogram plots of the corresponding sample conditions are presented in Fig. 10. In Fig. 10a, for the Ti-rich alloy in all the conditions, the transformation temperatures correspond to one-step M↔A phase transformation both while heating and cooling. While compared to the transformation temperatures of the AR sample, it is observed that, for the HPT sample, there is a slight decrease in M_f and A_s temperatures, whereas M_s and $A_f$temperatures increase. As a result, both while heating and cooling, there is a broadening of the temperature intervals in which the phase transformations take place. For the HPT sample after heat treatment at 300°C, designated as HPT+HTT300 in the plot, there is an increase in M_f and $A_s$temperatures, whereas M_s and A_f temperatures decrease. These results, both while heating and cooling, on narrowing of the temperature intervals where the phase transformations are taking place. After heat treatment at 350°C, designated as HPT+HTT350 in the plot, all the transformation temperatures increase and the phase transformation temperature intervals become narrower.

In Fig. 10b, for the Ni-rich alloy in the AR and HPT conditions, the transformation characteristics show a one-step M↔A phase transformation, both while heating and cooling. It is observed that for the HPT sample, the temperatures corresponding to both phase transformations are higher than those corresponding to the AR sample. However, both while heating and cooling, corresponding to M→A and A→M transformations, respectively, there is a narrowing of the transformation temperature intervals. For HPT+HTT300 sample, M_s decreases, A_s, and A_f increase considerably. M_f decreases to a value below the lower limit of the scanned temperature range. The dashed lines represent the trend of the variation of M_f. Further, R-phase transformations are present both while heating and cooling. On heat treatment at 350°C after the HPT processing, i.e., for Ni-rich HPT+HTT350, it is observed that all the transformation temperatures tend to increase.

AR samples and samples subjected to HPT of both alloys are scanned using XRD technique at different temperatures in the phase transformation temperature range. 3D view of the XRD profiles obtained while cooling and heating are presented in Fig. 11. Miller indices of the diffraction peaks emerging from the corresponding planes of the phases are marked on each peak. In Fig. 11a, for the Ti-rich Ni-Ti AR sample, it might be observed that the recording of the XRD pattern is started at 180°C, where austenite phase exists, followed by cooling and recording the spectra at different temperatures until the martensite transformation is complete, i.e., down to -40°C. Further, the sample is again heated to observe the transformation to austenite, i.e., up to 180°C to complete the thermal cycle. While cooling from 180°C to -40°C, the peak B2(1 1 0) corresponding to austenite (B2 – cubic structure) gradually disappears and peaks associated to martensite (B19' – monoclinic structure) gradually grow. The diffraction pattern obtained at -40°C, shows the peaks corresponding to martensite. A_s the temperature is increased from -40 to 180°C, the peak corresponding to (1 1 0) of austenite (B2 – structure) gradually grows and the peaks corresponding to B19' martensite gradually disappear. In Fig. 11b, for the Ti-rich Ni-Ti sample subjected to HPT, also M↔A phase transformation behavior is observed.

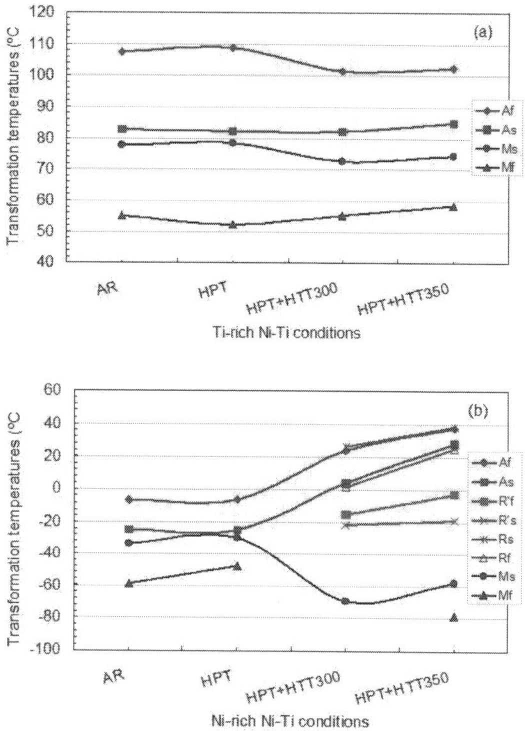

Figure 10. Phase transformation temperatures obtained from DSC plots of (a) Ti-rich and (b) Ni-rich Ni-Ti alloys in different conditions.

Fig. 11c shows the phase transformation behavior of Ni-rich Ni-Ti alloy in the AR condition. At 100°C, the sample is found to be in austenite (B2) phase. As the temperature is decreased down to -180°C, the intensity of the peak corresponding to B2(1 1 0) decreases. As the cooling progresses, the diffraction peaks corresponding to B19' martensite appear. On heating, the peaks related to B19' martensite disappear and the peak related to B2(1 1 0) appears again. Similar phase transformation behavior is observed for the Ni-rich sample after HPT (Fig. 11(d)). 3D layout of the XRD patterns obtained at selected temperatures during cooling, followed by heating for both Ti-rich and Ni-rich Ni-Ti alloys in HPT+HTT300C and HPT+HTT350C conditions were presented in a recent publication [47]. It is clearly observed that the diffraction peaks corresponding to intermediate R-phase are present for the Ti-rich and absent for the Ni-rich Ni-Ti alloys, both while cooling and heating. The result is in agreement with the transformation temperature profiles obtained by DSC thermogram analyses presented in the above Fig. 10.

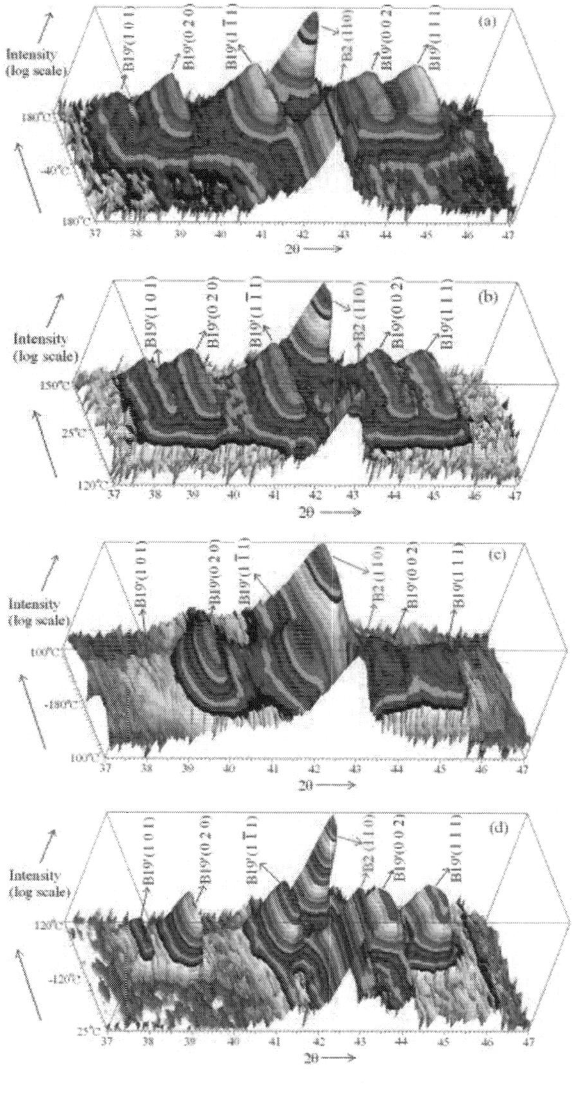

Figure 11. D box layout of the XRD profiles obtained during cooling and heating for Ti-rich Ni-Ti alloy in (a) AR and (b) HPT conditions, and Ni-rich Ni-Ti alloy in (c) AR and (d) HPT conditions.

The results show that for Ti-rich Ni-Ti alloy, after HPT, as well as following the heat treatments, there are no major changes in the phase transformation behavior. But, for Ni-rich Ni-Ti alloy, there is a slight change in the phase transformation behavior after HPT process, and the final heat treatments bring about very significant change, namely, the presence of intermediate R-phase transformation. In the present experiment, during the HPT process, a high speed of rotation of the piston (1,250 rpm) is involved. Initially, when the pressure torque is applied, a very intense and rapid plastic deformation takes place. This causes the specimen to get macroscopically distorted geometrical shape and eventually microscopic disorder. Owing to the process, the specimen gets heated up and might undergo a short duration annealing in the severely strained condition before cooling to room temperature. This situation may lead to accommodate several conflicting processes [46]. High speed of rotation during the HPT process might also trigger dynamic recrystallization. Depending on factors, such as the previous condition of the HPT specimen, strain accommodated, temperature attained, and magnitude of the time interval at which the specimen is at high temperature, different final microstructural states will be achieved in the specimen. On one hand, the intense deformation will distort the microstructure and long range order will be broken. On the other hand, the high temperature will have its influence on the recovery of the strains and formation of strain free submicrocrystals.

b. Equal channel angular pressing (ECAP) or Equal channel angular Extrusion (ECAE)

ECAP is an attractive processing technique for several reasons. Processing by ECAP can have a strong effect not only on the mechanical properties but also on the functional properties of materials [48]. However, for Ni-Ti SMA, it is difficult to apply ECAP at RT due to their low deformability and accordingly several reports have appeared describing the fabrication of ultrafine-grained alloys using ECAP at elevated temperatures [49]. The transformation behavior of TiNi alloy after ECAE process has been reported by **Zhenhua Li** *et al.*, [50] by using the experimental material, Ti-50.6at% Ni alloy rods, with a 25 mm diameter, after 850 °C hot forging and 500 °C annealing for 2 h. They concluded that during high temperature ECAE process, there was no dynamical re-crystallization but, most probably, there was dynamical recovery. Annealed for 5 min at 750 °C after two passes of ECAE, grains were refined and became even. After two passes of ECAE, transformation temperatures of the billet of TiNi alloy sharply decreased. Transformation temperature of the sample remarkably increased annealed for 2 h at 500 °C after two ECAE processes, similar to the one of TiNi alloy before ECAE process, which was related to Ni content in the matrix.

Effect of ECAP process on the microstructure and functional properties, such as recovery stress and maximum fully recoverable strain has been reported. The results show that the multipass ECAP of $Ni_{50.2}Ti_{49.8}$ alloy allows one to produce a uniform grain structure with predominantly high-angle grain boundaries with a grain size of about 200-300 nm. ECAP increases strength and insignificantly decreases plasticity as compared to the as-quenched state. The strength increases more than 50% with increasing number of passes; after ECAP using 12 passes. The functional properties of the $Ni_{50.2}Ti_{49.8}$alloy after ECAP are substantially improved. With increasing number of ECAP passes the maximum recovery stress rises to 1100 MPa and the degree of maximum fully recoverable strain increases to 9.2% [51].

8. CONCLUDING REMARKS

Phase transformations can be studied by using various characterization techniques, such as DSC, ER, Internal Friction (IF), dilatometry, XRD, and optical/electron microscopy [5, 14, 16-19, 41, 43, 52, 53]. Each of these techniques senses different physical phenomena and thus provides information concerning the changes of various physical parameters taking place during the phase transformations. Because of their distinctive nature, when these techniques are employed individually, only partial information about the phase transformation can be delivered.

DSC measures only the sum of all thermal events and, as a result, some important features may be ignored or the results are easily misinterpreted in the cases involving weak and/or complex (overlapping) transformations [5, 16, 18, 19]. ER is the structural sensitive property of a material and it reveals changes during crystallographic phase transformations. In fact, it is found to be more sensitive than DSC in detecting the phase transformations which occur in a narrow temperature range [19, 41]. Dilatometry is capable of sensing small volume changes during phase transformations. Only a limited number of publications report the use of dilatometry to study the phase transformations in Ni-Ti shape memory alloys [17, 19]. These methods have been widely accepted to detect the phase transformations in Ni-Ti SMAs. A combined approach of several characterization techniques would lead to the proper understanding of the phase transformations involved.

ACKNOWLEDGEMENT

The pluriannual financial support (by Fundação para a Ciência e a Tecnologia – Ministério da Educação e Ciência) of CENIMAT/I3N through the Strategic Project - LA 25 - 2011-2012 and the research project Smart Composites (PTDC/CTM/66380/2006) is gratefully acknowledged by KKM and FMBF. KKM gratefully acknowledges the fellowship under the scheme, 'Ciência 2007' with Ref. No. C2007-443-CENIMAT-6/Ciência2007.

REFERENCES

1. Effect of Thermal Cycling on the Transformation Temperatures of TiNi AlloysS. Miyazaki, Y. Igo, K. Otsuka. Acta Metall. 3419861 98620452051

2. Factors Affecting the Ms Temperature and its Control in Shape-Memory AlloysK. Otsuka, X. Ren. Mater. Sci. Forum 394-395 (2002

3. Lattice Transformations Related to Unique Mechanical EffectsJ. Perkins. Metall. Trans. 41973197327092721

4. Effect of Thermal Cycling on the R-phase and Martensitic Transformations in a Ti-Rich NiTi AlloyV. Pelosin, A. Riviere, Metall. Mater. Trans. A 291998199811751180

5. Effect of Thermal Cycling on R-phase Stability in a NiTi Shape Memory AlloyJ. Uchil, K.G. Kumura, K.K. Mahesh, Mater. Sci. Eng. A 332200220022528

6. Effect of Deformation and Thermal Treatment of NiTi Alloy on Transition SequenceMorawiec, D. Stroz, D. Chrobak. J. Phys. IV 519951995205209

7. Effect of Heat Treatment After Cold Working on the Phase Transformation in TiNi AlloyT. Todoroki, H. Tamura. Trans. J. Inst. Met. 28198719878394

8. A study on the wire drawing of TiNi shape memory alloys. S.K. Wu, H.C. Lin, Y.C. Yen. Materials Science and Engineering A 215 (1996) 113-119.

9. The effects of cold rolling on the martensitic transformation of an equiatomic TiNi alloyH.C. Lin, S.W. Wu, T.S. Chou, H.P. Kao. Acta Metall. Mater. 391991199120692080

10. Deformation of Martensite in a Polycrystalline Cu-Zn-Al AlloyK. Adachi, J. Perkins. Metall. Trans. 17A (1986

11. FEM Simulation of the Martensitic Transformation in NiTi AlloysS. Zhang, H. Braasch, P.G. McCormick. J Phys IV France 7 C51997 1997537542

12. In-situ High Temperature Texture Characterisation in NiTi Shape Memory Alloy Using Synchrotron RadiationA.S. Paula, K.K. Mahesh, F.M. Braz Fernandes, R.M.S. Martins, A.M.A. Cardoso, N. Schell. Materials Science Forum 495-497 (2005

13. Textural Modifications during Recovery in Ti-Rich Ni-Ti Shape Memory Alloy Subjected to Low Level of Cold Work ReductionA.S. Paula, K.K. Mahesh, N. Schell, F.M. Braz Fernandes, Materials Science Forum 636-637 (2010

14. Study of the textural evolution in Ti-rich NiTi using synchrotron RadiationA.S. Paula, J.H.P.G Canejo, K.K. Mahesh, R.J.C. Silva, F.M. Braz Fernandes, R.M.S. Martins, A.M.A. Cardoso, N. Schell. Nuclear Instruments and Methods in Physics Research Section B: Beam Interactions with Materials and Atoms 24620062006206210

15. Restoration Phenomena of Neutron-irradiated Ti-Ni Shape Memory AlloysT. Hoshiya, F. Takada, Y. Ichihashi. Materials Science and Engineering A 13019901990185191

16. Effect of Heat Treatment after Cold Working on the Phase Transformation in TiNi AlloyTsunehiko Todoraki, Hirokazu Tamura, Transactions of the Japan Institute of Metals 28(2) (1987

17. Thermal expansion in various phases of nitinol using TMAJ.Uchil, K.P.Mohanchandra, K.Ganesh Kumara, K.K.Mahesh, T.P.Murali. Physica B (UK) 27019991999289297

18. Thermomechanical behaviour of Ti-rich NiTi shape memory alloysA.S. Paula, K.K. Mahesh, C.M.L. dos Santos, F.M. Braz Fernandes, C.S. da Costa Viana. Materials Science and Engineering A (UK) 481-482, (2008

19. One- and Two-step Phase Transformation in Ti-rich NiTi Shape Memory AlloyA.S. Paula, K.K. Mahesh, C.M.L. dos Santos, J.P.H.G. Canejo, F.M. Braz Fernandes. International Journal of Applied Electromagnetics and Mechanics (Netherlands) 23200620062532

20. Microstructural evolution kinetics after plastic deformation of equiatomic Ti-Ni alloy during isothermal annealingsF. Khelfaoui, G. Thollet, G. Guenin. Mat. Sci. and Eng. A, 33820022002305312

21. Influence of Work Hardening and Heat Treatment on the Substructure and Deformation Behaviour of TiNi Shape Memory AlloysP. Filip, K. Mazanec. Scr. Met. et Mat. 321995199513751380

22. On the Ti2Ni precipitates and Guinier-Preston zones in Ti-rich Ti-Ni thin filmsJ.X. Zhang, M. Sato, A. Ishida. Acta Mater. 512003 200331213130

23. High-resolution electron microscopy studies on coherent plate precipitates and nanocrystals formed by low-temperature heat treatments of amorphous Ti-rich Ti-Ni thin filmsT. Kikuchi, K. Ogawa, S. Kajiwara, T. Matsunaga, S. Miyazaki, Y. Tomota. Philos. Mag. A 7819981998467489

24. Tratamentos Termomecanicos de Ligas do Sistema Ni-TiA.S. Paula, Ph.D. Thesis, FCT/UNL, Lisbon- Portugal, 2006

25. Experimental test for a possible isothermal martensitic transformation in a Ti-Ni alloyK. Otsuka, X. Ren, T. Takeda, Scripta Materialia (45)2 (2001

26. Effects of Heat Treatment on Mechanical Behaviour of Ti-Ni AlloysT. Saburi, T. Tatsumi, S. Nenno. Journal de Physique Colloques 43(C4) (December 1982Proceedings of the International Conference on Martensitic Transformations (ICOMAT-82) 261-266.

27. Neutron diffraction phase analysis during thermal cycling of a Ni-rich NiTi shape memory alloy using the Rietveld methodH. Sitepu, W.W. Schmahl, J. Khalil Allafi, G. Eggeler, A. Dlouhy, D.M. Tobbens, M. Tovar. Scripta Materialia, 4620022002543548

28. Effect of cooling process during heat treatment on martensitic transformation in Ni-Ti and Ni-Ti-Cr alloysJ. Uchil, K. Ganesh Kumara, K.K. Mahesh. Proc. International Conference on Martensitic Transformations (ICOMAT'02) Helsinkin University of Technology, Espoo, Dipoli, Finland, June 10-14, 2002, Edited by J. Pietikainen, O. Soderberg, J de Physique- IV, France 11220032003747750

29. Some aspects of the properties of NiTi shape memory alloyY. Liu, J. Van Humbeeck, R. Stalmans, L. Delaey. Journal of Alloys and Compounds 24719971997115121

30. Precipitation Processes in Near-Equiatomic TiNi Shape Memory AlloysM. Nishida, C.M. Wayman, T. Honma. Metallurgical Transactions A 17198619861505l515

31. Effects of hot rolling on the martensitic transformation of an Equiatomic Ti-Ni alloyH.C. Lin, S.K. We, Mater. Sci. Eng. A158199219928791

32. Effect of microstructure on the fatigue of hot-rolled and cold-drawn NiTi shape memory alloysK. Gall, J. Tyber, G. Wilkesanders, S. W. Robertson, R. O. Ritchie, H. J. Maier. Materials Science and Engineering A 48620082008389403

33. Microstructure and Thermo-Mechanical properties of NiTi shape memory alloysE. Hornbogen. Mat. Sci. Forum 455-456 (2004

34. Appearance of an Intermediate Phase with Thermal Cycling on the Transformation of NiTiH. Matsumoto. J. Mat. Sci. Letts. 101991 1991408410

35. Transformation Behavior of NiTi in Relation to Thermal Cycling and DeformationH. Matsumoto. Physica B19019931993115120

36. P Filip, K Mazanec, Influence of Cycling on the Reversible Martensitic-Transformation and Shape-Memory Phenomena in TiNi AlloysScripta Metallurgica et Materialia30199419946772

37. J Kwarciak, Z Lekston, H Morawiec, Effect of Thermal Cycling and Ti2Ni Precipitation on the Stability of the Ni-Ti AlloysJ. Mat. Sci. 221987198723412345

38. G Airoldi, B Rivolta, C Turco, Heats of Transformations as a Function of Thermal Cycling in NiTi Alloys, in Proc. Int. Conf. of Martensitic Transformations, JIM Ed., 19861986691696

39. D. A Hebda, S. R White, Effect of training conditions and extended thermal cycling on nitinol two-way shape memory behaviour. Smart Mater. Struct. 419951995298304

40. C. J De Araujo, M Morin, G Guenin, Estimation of internal stresses in shape memory wires during thermal cycling under constant load: A macromechanical approachJ. Int. Mat. Sys. & Structs. 1120002000516524

41. A. S Paula, K. K Mahesh, F.M. Braz Fernandes, Stability in Phase Transformation after Multiple Steps of Marforming in Ti-rich Ni-Ti Shape Memory AlloyJournal of Materials Engineering and Performance 2020112011771775

42. M Zehetbauer, R Grossinger, H Krenn, M Krystian, R Pippan, P Rogl, T Waitz, R Wurschum, Bulk Nanostructured Functional Materials By Severe Plastic DeformationAdv Eng Mater 1220102010692700

43. K. K Mahesh, F. M Braz, R. J. C Fernandes, Silva, G. Gurau. Phase transformation and structural study on the SPD NiTi alloys. Physics Procedia, 10201020102227

44. T Waitz, V Kazykhanov, H. P Karnthaler, Martensitic phase transformations in nanocrystalline NiTi studied by TEMActa Mat 5220042004137147

45. Alloy compositiondeformation temperature, pressure and post-deformation annealing effects in severely deformed Ti-Ni based shape memory alloys. S.D. Prokoshkin, i.Yu. Khmelevskaya, S.V. Dobatkin, I.B. Trubitsyna, E.V. Tatyanin,V.V. Stolyarov, E.A. Prokofiev Acta Mater 532005200527032714

46. K. K Mahesh, F. M Braz, Fernandes, G. Gurau. Stability of thermal induced phase transformations in the severely deformed equiatomic Ni-Ti alloysJ Mater Sci 472012201260056014

47. F. M Braz, K. K Fernandes, R. J. C Mahesh, C Silva, Gurau, G. Gurau. XRD study of the transformation characteristics of severely plastic deformed Ni-Ti SMAsPhys. Status Solidi C72010201013481350

48. R. Z Valiev, T. G Langdon, Principles of equal-channel angular pressing as a processing tool for grain refinementProg Mater Sci 5120062006881981

49. V. G Pushin, R.Z. Valiev The Nanostructured TiNi Shape-Memory Alloys: New Properties and Applications, in Proceedings of the Symposium on Interfacial Effects in Nanostructured Materials, Warsaw, Poland, 1418September 2002. Solid State. Phenom., 94, (2003

50. Z Li, G Xiang, X Cheng, Effects of ECAE process on microstructure and transformation behaviour of TiNi shape memory alloy. Materials and Design 2720062006324328

51. E Prokofyeva, D Gunderov, S Prokoshkin, R Valiev, Microstructure, mechanical and functional properties of NiTi alloys processed by ECAP technique, in Proceedings ESOMAT 20098TH European Symposium on Martensitic Transformations, Article nr 06028, 5 pgs.

52. D Stroz, Z Bojarski, J Ilczuk, Z Lekston, H Morawiec, Effect of thermal cycling on as-quenched and aged nickel-rich Ni-Ti alloyJournal of Materials Science2619911991l7411748

53. M Piao, K Otsuka, S Miyazaki, H Horikawa, Mechanism of the As temperature increase by pre-deformation in thermoelastic alloys.Mater. Trans JIM 3419931993919929

7

AB INITIO AND MONTE CARLO APPROACHES FOR THE MAGNETOCALORIC EFFECT IN CO- AND IN-DOPED NI-MN-GA HEUSLER ALLOYS

Vladimir Sokolovskiy [1,2], Anna Grünebohm [3], Vasiliy Buchelnikov [1] and Peter Entel [3,*]

[1]Chelyabinsk State University, Br. Kashirinykh 129, 454001 Chelyabinsk, Russia (V.S.)

[2]National University of Science and Technology "MISiS", Leninskiy pr. 4, 119049 Moscow, Russia

[3]University of Duisburg-Essen, Lotharstrasse 1, D-47048 Duisburg, Germany

ABSTRACT

The complex magnetic and structural properties of Co-doped Ni-Mn-Ga Heusler alloys have been investigated by using a combination of first-principles calculations and classical Monte Carlo simulations. We have restricted the investigations to systems with 0, 5 and 9 at% Co. *Ab initio* calculations show the presence of the ferrimagnetic order of austenite and martensite depending on the composition, where the excess Mn atoms on Ga sites show reversed spin configurations. Stable ferrimagnetic martensite is found for systems with 0 (5) at% Co and a c/a ratio of 1.31 (1.28), respectively, leading to a strong competition of ferro- and antiferro-magnetic exchange interactions between nearest neighbor Mn atoms. The Monte Carlo simulations with *ab initio* exchange coupling constants as input parameters allow one to discuss the behavior at finite temperatures and to determine magnetic transition temperatures. The Curie temperature of austenite is found to increase with Co, while the Curie temperature of martensite decreases with increasing Co content. This behavior can be attributed to the stronger Co-Mn, Mn-Mn and Mn-Ni exchange coupling constants in austenite compared to the corresponding ones in martensite. The crossover from a direct to inverse magnetocaloric effect in Ni-Mn-Ga due to the substitution of Ni by Co leads to the appearance of a "paramagnetic gap" in the martensitic phase. Doping with In increases the magnetic jump at the martensitic transition temperature. The simulated magnetic and magnetocaloric properties of Co- and In-doped Ni-Mn-Ga alloys are in good qualitative agreement with the available experimental data.

KEYWORDS

Heusler alloys; magnetocaloric effect; first-principles and Monte Carlo methods

1. INTRODUCTION

Among ferromagnetic (FM) materials having acceptable magnetocaloric properties, Ni-Mn-Ga Heusler alloys have been widely investigated by experimental and theoretical methods in view of their potential applications as intelligent functional materials [1–5]. The multifunctional properties usually appear across a martensitic transformation due to the strong coupling of crystal structure and magnetic order. The shape memory effect and magnetic-field-induced strains [6–10,12], superelasticity [13–15], magnetocaloric effect [16–19], as well as the spin glass and strain glass behavior [20–23] are found among these multifunctional properties. Generally, the tuning of both, structural and magnetic transition temperatures, can be useful to achieve better functional properties. At present, this optimization problem of Heusler alloys is being intensively investigated.

There is an extreme sensitivity of the crystal structure of the martensitic phase with respect to composition in off-stoichiometric $Ni_{2+x}Mn_{1-x}Ga$ or $Ni_2Mn_{1+x}Ga_{1-x}$ alloys [21,24–28], showing that in dependence of composition and thermo-mechanical treatments, the crystal structure of low-temperature martensite can be tetragonal or monoclinic, with five-layered modulation (5M) [29], orthorhombic or monoclinic, with seven-layered modulation (7M) [30,31], or non-modulated tetragonal (NM) [32,33].

Recently, the crossover between tetragonal and orthorhombic structures in Ni-Mn-Ga with a structural transition temperature close to room temperature has been reported by Richard *et al.* [28]. The authors have observed that the mixed region, where both tetragonal and orthorhombic structures can coexist, is near compositions with 50 at% Ni and Mn content between 28 and 30 at%; whereas the samples with high Mn content display the orthorhombic structure (7M), while compositions closer to 50 at% Ni show the tetragonal (5M) phase; see Figure 1. In addition, alloys like $Ni_{50}Mn_{30}Ga_{20}$($Ni_2Mn_{1.2}Ga_{0.8}$) exhibit a large magneto-crystalline anisotropy, which makes the alloy suitable for shape memory applications at room temperature, due to the exhibition of favorable structural ordering and twin-boundary motion.

A way to significantly improve the magnetic and mechanical properties of Ni-Mn-Ga alloys is related to adding a fourth element with the partial substitution of Ga, Ni or Mn. It should be noted that the fourth element should have a covalent radius and a valence electron concentration close to the parent elements of the alloy in order to avoid too large volume or strain effects. An interesting doping element is a transition metal element, like Co. In general, if the doping element decreases the overall valence electron concentration e/a, we expect that T_m will decrease, too. Therefore, when Co ($3d^74s^2$) substitutes for Ni ($3d^84s^2$), e/a decreases, but is increased in the case of Co replacing Mn ($3d^54s^2$) or Ga ($4s^24p^1$). If using Co, it is expected that the Curie temperature will increase because of the strong ferromagnetic nature of Co.

During the past decade, a series of investigations of the magnetic and structural properties of Co-doped Ni-rich and Mn-rich Ni-Mn-Ga Heusler alloys has been reported by different groups regarding the substitution of Co for Ni, Mn, as well as Ga. Since, in this paper, we present the results of *ab initio* calculations and Monte Carlo simulations of substitutional processes (mainly Co for Ni), we have gathered in Table 1 the investigations from the literature (without claiming to have a complete list of all investigations performed so far) [19,34–50].

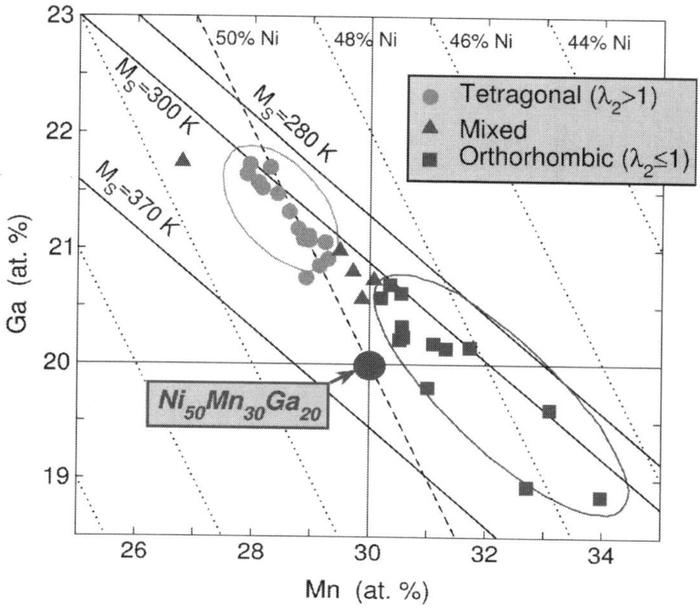

Figure 1. Phase diagram of Mn-rich Ni-Mn-Ga alloys showing the region where tetragonal and orthorhombic structures may coexist at room temperatures (denoted as "mixed" [28]). The dashed line indicates compositions with 50% Ni. The middle eigenvalue of the transformation matrix is marked in the panel and has been derived from measurements of the lattice constant of austenite and martensite [28]. An eigenvalue close to one indicates a possible coherent lattice invariant martensite-austenite interface.

In general, it has been established that in Ni-Mn-Ga-Co compounds, the martensitic transformation temperature is found to increase with the substitution of Mn or Ga by Co and to decrease when replacing Ni by Co. Hence, the variation of the martensitic transition temperature follows the general trend that T_m is proportional to the average valence electron number per atom (e/a) ratio. Concerning the Curie temperature, T_C increases with the substitution of Co for Ni and shows a weak effect with non-monotonous behavior when Mn or Ga is replaced by Co. Interestingly, in samples with the substitution of Ga by Co, T_C is almost independent of the doping concentration when the FM transition takes place in austenite, while it decreases with increasing x when the FM order occurs within the martensite [38,46]. With respect to the critical value of Co content in the Ni sublattice for which the martensitic phase disappears, Cong *et al.* [35] have observed an abrupt decrease ofT_m when the Co exceeds 6% in Ni-rich $Ni_{2+x-y}Co_yMnGa_{1-x}$ ($x = 0.12$, $0 \leq y \leq 0.56$), *i.e.*, T_m decreases drastically from 375 K to 236 K when the Co content changes from $y = 0.24$ to 0.32. In the alloys with $y \geq 0.4$, no martensitic transformation was observed in the temperature range 188–573 K in the DSC measurements [35]. A recent study of Ni-rich $Ni_{2-y}Co_yMn_{1+x}Ga_x$ (x= 0.16, $0 \leq y \leq 0.4$) performed by Pushin *et al.* [44] has shown that upon Co doping (up to $y = 0.4$), the structural transition temperature is found to decrease from room temperature to 6–10 K. Another interesting point is related to the possibility of inducing a paramagnetic gap in $Ni_{2-y}Co_yMn_{1+x}Ga_{1-x}$ ($0.2 \leq x \leq 0.28$, $0 \leq y \leq 0.36$) [41,43] between weakly magnetically ordered martensite and FM austenite. As a result, a reverse magnetostructural phase transformation appears, which is accompanied by a positive change of ΔS (called inverse MCE). In addition, the inverse MCE across the magnetostructural transition from FM austenite to PM martensite was also found in quinary $Ni_{2-y}Co_yMn_{1+x}Ga_{1-x-z}In_z$ ($0.26 \leq x \leq 0.28$, $0.35 \leq y \leq 0.368$, $0.08 \leq z \leq 0.132$) [19,48,49]. With respect to the sequence of phase transitions in Co doped in $Ni_{2-y}Co_yMn_{1+x}Ga_{1-x}$, Segui *et al.* [50] have recently stated that the transformation sequence may depend on the quenched-atomic disorder of austenite *i.e.*, poorly ordered alloys ($L2_1$ austenite) with a high e/a value show a $A_{PM} \Leftrightarrow M_{PM} \Leftrightarrow M_{FM}$ sequence, whereas disordered alloys (B2 austenite) display the $A_{PM} \Leftrightarrow A_{FM} \Leftrightarrow M_{PM} \Leftrightarrow M_{FM}$ sequence; where $A(M)$ denotes austenite (martensite), respectively. On the other hand, for samples with a low e/a value, the transformation sequence changes from$A_{PM} \Leftrightarrow A_{FM} \Leftrightarrow M_{PM} \Leftrightarrow M_{FM}$ to $A_{PM} \Leftrightarrow A_{FM} \Leftrightarrow M_{FM}$.

In order to understand the influence of Co addition on the magnetic and structural properties of Ni-Mn-Ga more profoundly, several theoretical studies have been performed using model and first-principles calculations in the framework of density functional theory (DFT) [46,51–54]. Recently, Kanomata *et al.* [46] have analyzed the phase diagram of $Ni_2MnGa_{1-y}Co_y$ ($0 \leq y \leq 0.2$) using a Landau-type free energy expansion in strain and magnetic order parameters. They observed that for optimal model parameters, the theoretical phase diagram reproduces the main features of the experimental phase diagram. On the basis of DFT calculations, Entel *et al.* [51] have studied the effect of Ga substitution by Co (3 at%) on the physical properties of Ni_2MnGa, finding that the addition of Co causes an increase of magnetic moments and an increase of the energy difference between austenite and martensite phases. The estimated values of T_C and T_m were found to increase by 80 and 140 K, respectively.

Table 1. The variety of Co-doped Ni-Mn-Ga Heusler alloys that have been recently investigated.

Substitution of Co for Ni	Substitution of Co for Mn	Substitution of Co for Ga
$Ni_{2+x-y}Co_yMn_{1-x}Ga$	$Ni_2Mn_{1+x-y}Co_yGa_{1-x}$	$Ni_2Mn_{1+x}Ga_{1-x-y}Co_y$
$x = 0.16, 0 \leq y \leq 0.09$ [34]	$x = 0.16, 0.0 \leq y \leq 0.26$ [40]	$x = 0, 0 \leq y \leq 0.25$ [38]
$Ni_{2+x-y}Co_yMnGa_{1-x}$	$Ni_{2+x}Mn_{1-y}Co_yGa_{1-x}$	$x = 0, 0 \leq y \leq 0.2$ [46]
$x = 0.12, 0 \leq y \leq 0.56$ [35]	$x = 0.24, 0.16 \leq y \leq 0.32$ [36]	$x = 0.16, y = 0.088$ [37]
$x = 0.24, 0 \leq y \leq 0.32$ [36]	$Ni_{2+x}Mn_{1-x-y}Co_yGa$	$x = 0.16, 0 \leq y \leq 0.17$ [47]
$Ni_{2-y}Co_yMn_{1+x}Ga_{1-x}$	$x = 0, 0 \leq y \leq 0.25$ [45]	**T_m increases,**
$x = 0, 0 \leq y \leq 0.2$ [38]	$x = 0, 0.03 \leq y \leq 0.2$ [38]	**T_C weakly changes**
$x = 0.16, 0 \leq y \leq 0.16$ [40]	$x = 0.16, 0.04 \leq y \leq 0.14$ [34]	
$x = 0.16, 0 \leq y \leq 0.13$ [42]	**T_m increases,**	
$x = 0.16, 0 \leq y \leq 0.4$ [44]	**T_C weakly changes**	
$x = 0.24, 0.20 \leq y \leq 0.24$ [39]		
$0 \leq x \leq 0.11, 0.068 \leq x \leq 0.232$ [37]		
$0.2 \leq x \leq 0.28, 0 \leq y \leq 0.36$ [19,41]		
$0.2 \leq x \leq 0.28, 012 \leq y \leq 0.32$ [43]		
T_m decreases, T_C increases		

The effect of Co content on the magnetic property and phase stability of $Ni_{2-y}Co_yMnGa$ ($0 \le y \le 0.75$) alloys has also been investigated by Chang-Long *et al.* [52]. They show that both the phase stability of austenite and the Curie temperature increase with increasing Co content. Similar investigations of the physical properties of $Ni_{2-y}Co_yMnGa$ ($0 \le y \le 1$) have been performed by Bai *et al.* [53]. They observed that the lattice parameter a of austenite decreases, while for martensite, $a$increases and c decreases with increasing Co content. Besides, the Curie temperature increases with Co content. Detailed *ab initio* investigations of Co- and Cu-doped Ni_2MnGa along the tetragonal deformation path have been recently presented by Zeleny *et al.* [54] with substitution of Co for Ni and Cu for Mn or Ga with up to 12.5%. These simulations show that Co doping has a much stronger effect on the equilibrium (c/a) value compared to Cu. Moreover, the increase of Co content causes the equilibrium c/a ratio to decrease from 1.25 (0% Co) to 1.16 (7.5%), and a further increase of Co concentration up to 12.5% destabilizes the martensitic phase. The theoretical investigations widely confirm the experimental trends and properties of the Ni-Mn-Ga-Co alloys.

In this paper, we extend the zero-temperature first-principles calculations of Co-doped Ni-Mn-Ga within the DFT scheme to finite-temperature Monte Carlo (MC) simulations, where exchange coupling constants, magnetic moments, and anisotropy energy are taken from *ab initio* calculations. This allows us to simulate the temperature dependence of magnetic and thermodynamic properties, as well as the magnetocaloric effect of Ni-Co-Mn-Ga across a magnetostructural phase transformation.

2. CONSTRUCTION OF THE MODEL HAMILTONIAN

For the finite-temperature MC simulations of Co-doped Ni-Mn-Ga alloys, we employ the extended Potts–Blume–Emery–Griffiths Hamiltonian, which allows one to describe a system of interacting magnetic moments on a polycrystalline lattice. Each atom of the lattice is characterized by magnetic S_i and structural σ_i degrees of freedoms. Since the magnetic moment of Ga is about several orders of magnitude smaller than the magnetic moments of Mn, Ni, and Co, we consider only the structural degree of freedom for Ga. We would like to point out that in case of non-stoichiometric $Ni_{50}Mn_{25+x}Z_{25-x}$ (Z = Ga, In, Sn, Sb) Heusler alloys, there are Mn magnetic moments interacting antiferromagnetically (AFM) with each other. The first Mn atom, which is referred to as Mn_1, occupies the original Mn sublattice, while the second Mn atom (Mn_2) is located on the Ga sublattice. *Ab initio* calculations reveal strong AFM Mn_1-Mn_2 interactions, because of the shorter distance between Mn_1-Mn_2 ($d = 0.5a_0$) compared to Mn_1-Mn_1 and Mn_2-Mn_2 distances (d=2√/2a0) [55–60,66]. Here, a_0 is the lattice parameter of the cubic $L2_1$.

In order to describe approximately the effect of polycrystallinity (or multidomain magnetic states), we divided the three-dimensional simulation cell into 16 domains with random initial spin configurations. For the q-state Potts model, the spin configurations correspond to discrete spin states $S_i \in \{1, 2, ..., q\}$, where the maximum number of possible q states depends on a total spin moment S of an atom defined by the $2S + 1$ multiplet. In each magnetic domain, all atoms have the same configuration as the variable S_k, which defines the 'easy axis' and characterizes the action of magnetic anisotropy in each domain. Similar to the multi-domain Heisenberg model, all spins will be aligned with one another pointing in the same direction as enforced by the magnetic anisotropy. Hence, for the Heisenberg and Potts model, different domain magnetizations will approximately lead to zero total magnetization in a zero or low external magnetic field.

In this work, our main concern is the q-state Potts model in combination with the three state Blume–Emery–Griffiths (BEG) model, which was previously applied to Heusler systems [55,60–65]. Note that the Potts model is suitable for investigations of first-order phase transitions, as well as recrystallization problems, such as grain growth kinetics, grain size distribution and their topologies, *etc.* We would like to note that most of the formulas have already been published (see, [55,60–65]), but for the sake of completeness and discussion, we briefly outline here the model Hamiltonian. In the Potts HamiltonianEquation (1), the magnetic exchange interactions, magnetic anisotropy, and externally magnetic fields are specified by:

$$\mathcal{H}_{mag} = -\sum_{\langle ij \rangle} J_{ij}^{m} \delta_{S_i,S_j} + K_{ani} \sum_{i} \delta_{S_i,S_k} \mu_i^2 - g\mu_B H_{ext} \sum_{i} \delta_{S_i,S_g} \mu_i. \tag{1}$$

where J_{ij} are the exchange coupling constants between magnetic moments μ_i at sites i and j of cubic and tetragonal Heusler lattices. S_i is the Potts' spin at site i, which can take on q integer values depending on the total spin moment S of an atom. The spin moment of Mn is $S = 5/2$; hence, we associate the $2S + 1$ possible spin projections (−5/2, −3/2, −1/2, 1/2, 3/2 and 5/2) with the $q_{Mn} = 1 ... 6$ states. Likewise, we assume $S = 1$ (3/2) for Ni (Co), respectively, with the possible $q_{Ni} = 1 ... 3$ ($q_{Co} = 1$

...4) states. K_{ani} is the magnetic anisotropic constant. H_{ext} is the external magnetic field. μ_B is Bohr's magneton, and g is the Landé factor. The values of J_{ij} and magnetic moments μ_i are obtained from *ab initio* calculations for austenite and martensite separately. The first Kronecker symbol, δ_{S_i,S_j}, restricts the spin-spin interactions to the interactions between the same $S_i$states for Mn, Co and Ni. The coupling of the spin S_i to the magnetic anisotropy energy K_{ani} is specified by the second Kronecker symbol δ_{S_i,S_k}. Here, all spins in a domain have then the same q state as S_k; S_k may be chosen to be different from domain to domain. The coupling to the external magnetic field is specified in the third term by S_g, where we align all spins by choosing $S_g = 1$ ($H_{ext} \geq 0$). $\langle i, j\rangle$ denotes here, for the magnetic interactions, a sum over neighbors up to the sixth coordination shell in the Heusler lattice.

The structural part is described by the degenerated three-state Blume–Emery–Griffiths (BEG) model, which allows for a structural transformation from the cubic (austenitic) phase to the tetragonal (martensitic) phase,

$$\begin{aligned} \mathcal{H}_{el} &= -\sum_{\langle ij\rangle} \sigma_i\sigma_j \left\{ J + g\mu_B H_{ext} \left(U_1 \sum_i \delta_{S_i,S_g} + U_2 g\mu_B H_{ext} \right) \right\} \\ &\quad - K \sum_{\langle ij\rangle} \left(1-\sigma_i^2\right)\left(1-\sigma_j^2\right) - k_B T \ln(p) \sum_i \left(1-\sigma_i^2\right), \end{aligned} \qquad (2)$$

where J and K are the structural coupling constants for the tetragonal and cubic states, respectively. The variable σ_i defines the deformation state at each lattice site where $\sigma_i = \pm 1$ and $\sigma_i = 0$ specify the tetragonal and cubic phases, respectively. p is the degeneracy factor of the cubic phase that characterizes the number of structural variants. U_1 and U_2 are dimensionless magnetoelastic interaction constants. T is the temperature, and k_B is the Boltzmann constant. The summation $\langle i, j\rangle$ is taken over nearest neighbor pairs only. We would like to point out that in previous work [60,62–64], we have used a slightly different term $\mathcal{H}_{el}$ with respect to the action of the external magnetic field, where a shift of the structural transition temperature by the applied magnetic field had been modeled using the previous $\mathcal{H}_{el}$ Hamiltonian (using only the U_1 parameter) for not too large magnetic fields. However, this is insufficient to model the martensitic transformation in the case of large magnetic fields. Therefore, we added to Equation (2) the second and third terms with dimensionless magnetoelastic constants U_1 and U_2. Somewhat similar terms ($\eta_4 e^2 m\mu_B H$ and $\eta_3 e^2(\mu_B H)^2$) have been used in a Landau functional theory of elastic and magnetic interactions in [67].

The coupling between the magnetic and structural subsystems is defined in terms of the magnetostructural interaction as:

$$\mathcal{H}_{int} = 2\sum_{\langle ij\rangle} U_{ij}\,\delta_{S_i,S_j}\mu_i\mu_j\left\{\left(\tfrac{1}{2}-\sigma_i^2\right)\left(\tfrac{1}{2}-\sigma_j^2\right)-\tfrac{1}{4}\right)\right\}, \tag{3}$$

where U_{ij} are magnetoelastic interaction parameters.

The total Hamiltonian, which is used in the Monte Carlo simulations, is then the sum of all terms,

$$\mathcal{H} = \mathcal{H}_{mag} + \mathcal{H}_{el} + \mathcal{H}_{int}. \tag{4}$$

For a halfway realistic description of low-isofield magnetization curves of the magnetocaloric materials as a function of temperature in polycrystalline Heusler alloys, we need to specify intra-domain, as well as inter-domain interactions. This issue is somewhat ambiguous, because magnetic domains are separated by domain walls in ferromagnetic materials by several tens of nm in width [68], depending on the ratio of J_{ij} and and K_{ani}. Furthermore, one would have to take into account the magnetic dipolar interaction in a finite system in analogy to the interaction between magnetic nanoparticles. On the other hand, the number of atoms in the simulation cell used in the MC simulations is of the order of a few thousand of atoms giving rise to a length scale of *ca.* 50 nm, which is not much larger than the typical width of magnetic domain walls in metals. Therefore, for the sake of simplicity and computational time saving, we assume that spins belonging to neighboring domains do not interact in a zero magnetic field. Although this is a very crude approximation, it would be helpful to perform the MC simulations with the Potts model. However, for a finite magnetic field, we want to account for the competition of external and anisotropy fields. We use a very simple procedure by choosing a random number $r \in [0, 1]$ and compare it with the probability containing the ratio of anisotropy and Zeeman energy,

$$\mathcal{W}_d^{(i)} = \min\left\{1,\ \exp\left(-\frac{\mu_i^2|K_{ani}|}{g|\mu_i|\mu_B|H_{ext}|}\right)\right\}. \tag{5}$$

Such that for $r < \mathcal{W}_d$, we take into account both the intra-domain and inter-domain exchange interactions using the J_{ij} coupling constants; otherwise, only the intra-domain interactions are taken into account. Obviously, it follows from Equation (5) that in the case of a magnetic field, when the Zeeman energy is smaller than the magnetic anisotropy energy, W_d becomes small and, effectively, a limited number of atoms will participate in the inter-domain exchange interactions, resulting in a small total magnetization. In the case of a high magnetic field, when the field energy is much larger than the anisotropy energy, W_d approaches one, and all spins will participate in the inter-domain interactions aligning spins along the applied magnetic field. This leads to the disappearance of magnetic domains and to a maximum value of magnetization as a single crystal case.

Magnetic (m) and strain (ϵ) order parameters of our extended Potts-BEG model are given by:

$$m_a = \frac{1}{N_a}\left(\frac{q_a N^a_{max} - N_a}{q_a - 1}\right), \epsilon = \frac{1}{N}\sum_i \sigma_i. \quad (6)$$

Here, N_a is the number of Ni, Co or $Mn_{1(2)}$ atoms, Namax is the maximum number of identical magnetic states on the lattice, respectively, and q_a is the number of Ni, Co or $Mn_{1(2)}$magnetic states. In the case of $\epsilon = 0$, the austenitic phase is stable, while $\epsilon = 1$ corresponds to the martensitic phase with one martensitic variant.

In principle, in Ni-(Co)-Mn-Ga Heusler alloys, cubic austenite (A) and tetragonal martensite (M) with different magnetic moments can coexist across the martensitic phase transformation. Therefore, in order to estimate the total magnetization, we take into account the volume fractions of both phases, which are related through the formula:

$$f^{\mathrm{A}}(\%) + f^{\mathrm{M}}(\%) = 100\%, \quad (7)$$

where $f^{A(M)}$ are the volume fractions of austenite and martensite, respectively. Finally, the total magnetization can be expressed as:

$$M = M^{\mathrm{A}} f^{\mathrm{A}} + M^{\mathrm{M}} f^{\mathrm{M}}, \quad (8)$$

with:

$$M^{\mathrm{A(M)}} = 2\mu_{\mathrm{Ni}}^{\mathrm{A(M)}} m_{\mathrm{Ni}}(1-y) + y\mu_{\mathrm{Co}}^{\mathrm{A(M)}} m_{\mathrm{Co}} + \mu_{\mathrm{Mn_1}}^{\mathrm{A(M)}} m_{\mathrm{Mn_1}} + x\mu_{\mathrm{Mn_2}}^{\mathrm{A(M)}} m_{\mathrm{Mn_2}} + (1-x)\mu_{\mathrm{Ga}}^{\mathrm{A(M)}} m_{\mathrm{Ga}}, \quad (9)$$

$$f^{\mathrm{A}} = \frac{1}{N}\sum_i \sigma_i^{(0)}, \quad f^{\mathrm{M}} = \frac{1}{N}\sum_i \left(\sigma_i^{(+1)} + \sigma_i^{(-1)}\right). \tag{10}$$

Since the magnetic moment of Ga is much smaller than magnetic moments of Mn, Ni and Co, we can neglect the last term in Equation (9). In the present paper, we consider two types of martensitic variants, which can be characterized by the elongation ($\sigma_i = +1$) and contraction ($\sigma_i = -1$) of the cubic lattice ($\sigma_i = 0$) along one of the Cartesian directions during a temperature change.

In order to discuss the MCE of Ni-Mn-Ga-(Co) alloys, we must calculate the temperature dependence of the magnetic part of specific heat C_{mag} and entropy S_{mag} and the total specific heat $C = C_{mag} + C_{lat}$ with magnetic and lattice contributions. We neglect the electronic part of the specific heat. For the lattice heat, we have used the standard Debye approximation [62].

$$C_{mag}(T, H_{ext}) = \frac{1}{k_B T^2}\left[\langle\mathcal{H}^2\rangle - \langle\mathcal{H}\rangle^2\right], \tag{11}$$

$$S_{mag}(T, H_{ext}) = \int_{T_1}^{T_2} dT \, \frac{C_{mag}(T, H_{ext})}{T}. \tag{12}$$

The isothermal entropy and adiabatic temperature changes with varying external magnetic field can be obtained from [60,62–65],

$$\Delta S_{mag}(T, H_{ext}) = S_{mag}(T, H_{ext}) - S_{mag}(T, 0), \tag{13}$$

$$\Delta T_{ad}(T, H_{ext}) = -T\,\frac{\Delta S_{mag}(T, H_{ext})}{C(T, H_{ext})}. \tag{14}$$

Here, $S_{mag}(T,H_{ext})$ and $S_{mag}(T, 0)$ denote the magnetic entropy in the presence of magnetic field H_{ext} and in a zero field, respectively.

3. COMPUTATIONAL DETAILS

The investigations of magnetic and magnetocaloric properties of $Ni_{50-y}Co_yMn_{30}Ga_{20}$ (y = 0 and 5 at%) and $Ni_{41}Co_9Mn_{32}Ga_{16}In_2$ Heusler alloys have been carried out using electronic structure *ab initio* calculations and Monte Carlo simulations. In order to evaluate the magnetic exchange coupling constants, magnetic moments, and equilibrium magnetic reference states in Co- and In-doped alloys, we first performed *ab initio* calculations for the cubic $L2_1$ structure with lattice parameter a_0 = 5.85 Å and the tetragonal $L1_0$ structure with different tetragonal ratios, c/a, assuming that the volume of the unit cell does not practically change with structural distortion. The equilibrium lattice parameter a_0 was taken from our recent *ab initio* calculations [66] using the Vienna *ab initio* simulation package (VASP). In that work, we have done equilibrium energy calculations with tetragonal distortion c/a using 16-atom supercell compositions $Ni_8Mn_5Ga_3$ and $Ni_7Co_1Mn_5Ga_3$ corresponding to $Ni_{50}Mn_{31.25}Ga_{18.75}$ and $Ni_{43.75}Co_{6.25}Mn_{31.25}Ga_{18.75}$, respectively. Calculations have been carried out for two spin configurations referred to as "ferro" (*i.e.*, all spins are parallel) and "ferri" (the spin of Mn_2 on the Ga-lattice is reversed). It has been found that the "ferri" solution is energetically favorable for both alloys, and stable martensite is achieved for c/a = 1.31 (c/a = 1.28) for $Ni_8Mn_5Ga_3$ ($Ni_7Co_1Mn_5Ga_3$), respectively [66]. For calculations of the coupling constants, J_{ij}, we used lattice parameters and magnetic structures for $L2_1$ austenite and tetragonal $L1_0$ martensites corresponding to c/a = 1.31 (c/a = 1.28) for $Ni_{50}Mn_{30}Ga_{20}$ ($Ni_{45}Co_5Mn_{30}Ga_{20}$ and $Ni_{41}Co_9Mn_{32}Ga_{16}In_2$), respectively.

The calculation of exchange coupling constants has been performed for "ferro" and "ferri" configurations in austenite and martensite using the SPR-KKR package [69,70], where the Heisenberg coupling constants J_{ij} are determinated by employing Liechtenstein's method [71] and the spin-polarized scalar-relativistic mode and atomic sphere approximation (ASA). Chemical disorder in the off-stoichiometric alloys was treated in the framework of the coherent potential approximation (CPA). The angular momentum expansion was up to l_{max} = 3. In the self-consistency cycle and in calculation of J_{ij}, 4,495 *k*points were generated by a *k*-mesh grid of {57, 57, 57}. All calculations were converged to 0.01 mRy of total energy. For the exchange correlation energy, we used the general gradient approximation (GGA) in the form of Perdew, Burke and Ernzerhoff [72].

Subsequently, classical MC simulations were carried out in order to obtain the magnetic and magnetocaloric properties of Co- and In-doped Ni-Mn-Ga alloys as a function of temperature and magnetic field. The magnetic moments and exchange couplings obtained from *ab initio* calculations are used as input data. The MC simulations have been performed for a large enough simulation cell with periodic boundary conditions using the Metropolis algorithm [62]. Changes of the independent variables q_{Ni}, q_{Co}, $q_{Mn1(2)}$ and σ_i are accepted or rejected according to the single-site transition probability $W = \min\{1, \exp(-\Delta\mathcal{H}/k_BT)\}$.

As a time unit, we used one MC step consisting of N attempts to change the variables q_{Ni}, q_{Co}, $q_{Mn1(2)}$ and σ_i. We would like to remind the reader that in the case of multidomain structures, the procedure of acceptation of long-range interactions between spins from neighboring domains with probability $\mathcal{W}_d$ (see, Equation (5)) is performed for each MC step before the procedure of acceptance or rejection of independent variables q_{Ni}, q_{Co}, $q_{Mn1(2)}$ and σ_i. A uniform random number r is chosen from the interval [0, 1] and compared with the probability $\mathcal{W}_d$: if $r < \mathcal{W}_d$, then interactions between randomly chosen atoms i and its neighbors from other domains are taken into account. Otherwise, if $r > \mathcal{W}_d$, the interactions between atom i and its neighbors located in the same domain are taken into account. For each temperature the properties (internal energy of the system $\langle\mathcal{H}\rangle$ and magnetic $\langle m\rangle$ and structural $\langle\epsilon\rangle$ order parameters) were analyzed allowing $5 \cdot 10^5$ MC steps and 10^4 thermalization steps. For simplicity, we consider only two martensitic variants, *i.e.*, the degeneracy factor $p = 2$. Values of spin states (*i.e.*, the q_{Ni}, q_{Co}, $q_{Mn1(2)}$variables) have been chosen randomly by using a number r with $0 \leq r \leq 1$ and fixing q_{Ni}, q_{Co}, $q_{Mn1(2)}$ according to the scheme: if $0 \leq r \leq l/3$, then $q_{Ni} = l$, $l = 1, 2, 3$, if $0 \leq r \leq k/4$, then $q_{Co} = k$, $k = 1, \ldots, 4$, and if $0 \leq r \leq n/6$, then $q_{Mn1(2)} = n$, $n = 1, \ldots, 6$.

The simulations have been done for $N = L^3$ lattice sites, where L is the number of cubic unit cells with lattice parameter a_0; here, $L = 6$ has been taken. Thus, for stoichiometric Ni_2MnGa, we have used a simulation cell containing 1,098 Mn, 1,099 Ga and 1,728 Ni atoms. For off-stoichiometric $Ni_{2-y}Co_yMn_{1+x}Ga_{1-x}$ alloys, we assume that fractions of Mn_2 (x) and Co (y) are randomly distributed on the Ga and Ni sublattices, respectively, according to the alloy composition. In order to create the multi-domain model lattice, we divided the simulation cell into 16 sublattices (domain blocks) with initial random q-Potts variables (*i.e.*, the overall spin configuration in each domain is different). For example, for stoichiometric Ni_2MnGa, this amounts to a total of 68 Mn and 108 Ni atoms in each of the 16 domains. In the case of a zero magnetic field, we assume that only intra-domain interactions are present, while for a finite magnetic field, inter-domain interactions are switched using the statistical factor $\mathcal{W}_d$ for the competition of magnetic anisotropy and Zeeman energies.

We would like to note that part of the parameters used in the simulations, like magnetic exchange coupling constants $J_{ij}$and magnetic moments μ_i, have been taken from *ab initio* calculations, while the other parameters, like structural exchange couplings J and K, magnetoelastic interaction constants U_{ij}, U_1 and U_2 and anisotropy constants K_{ani}, have been used as fit parameters in order to reproduce the experimental martensitic transformation temperature and magnetization behavior in different magnetic fields. It is worth noting that the model values of anisotropy and magnetoelastic interaction constants are close to values obtained from experiments and *ab initio* calculations for Ni-Mn-Ga [73–75]. The model parameters listed inTable 2 and 3 allow, therefore, reasonable simulations of the MCE. We would like to add that in the case of $Ni_{41}Co_9Mn_{32}Ga_{16}In_2$, the "ferro" spin configuration in austenite is more stable compared with the "ferri" spin configuration, because an increase in Co content leads to an enhancement of ferromagnetism, due to the strong FM exchange interactions between Mn and Co.

Table 2. Model parameters (in meV) for Co- and In-doped Ni-Mn-Ga alloys. Parameters U_1 and U_2 are the dimensionless constants.

Composition	J	K	U_{ij}^{A}	U_{ij}^{M}	U_1	U_2	K_{ani}^{A}	K_{ani}^{M}
$Ni_{50}Mn_{30}Ga_{20}$	6.6	0.5	0.5	3	0.1	0.1	0.00001	0.01
$Ni_{45}Co_5Mn_{30}Ga_{20}$	6.2	0.5	2	0.5	−0.1	−0.1	0.00002	0.02
$Ni_{41}Co_9Mn_{32}Ga_{16}In_2$	6.3	0.5	0.4	0.2	−0.2	−0.8	0.00005	0.05

4. COMPUTATIONAL RESULTS AND DISCUSSION

4.1. Results of Ab Initio Calculations

In order to complete the magnetic characterization of $Ni_{50-y}Co_yMn_{30}Ga_{20}$ (y = 0 and 5 at%) and $Ni_{41}Co_9Mn_{32}Ga_{16}In_2$Heusler alloys, we now consider the J_{ij} in Figure 2 and the electronic density of states (DOS) in Figure 3 of $L2_1$ and $L1_0$Heusler structures for the two different spin configurations labeled "ferri" and "ferro" solutions. The corresponding J_{ij} have been split into $Mn_{1(2)}$-X (X = $Mn_{1(2)}$, $Mn_{2(1)}$ Ni, and Co) and Co-Y (Y = Co and Ni), respectively, where the index "1" refers to Mn atoms on the Mn sublattice and index "2" to Mn atoms on the Ga sublattice. For the J_{ij} calculations, the SPR-KKR code and CPA [69,70] have been used.

Table 3. Magnetic moments (in μ_B) of Co- and In-doped Ni-Mn-Ga alloys for different structures obtained from *ab initio* calculations using the SPR-KKR package [69].

Composition	μ_{Mn_1}	μ_{Mn_2}	μ_{Ni}	μ_{Co}	μ_{Ga}	μ_{In}	μ_{tot}
$Ni_{50}Mn_{30}Ga_{20}$ $(c/a = 1)$ "ferri"	3.547	−3.746	0.242		−0.077		3.219
$Ni_{50}Mn_{30}Ga_{20}$ $(c/a = 1.31)$ "ferri"	3.485	−3.61	0.274		−0.075		3.25
$Ni_{45}Co_5Mn_{30}Ga_{20}$ $(c/a = 1)$ "ferri"	3.512	−3.74	0.259	0.938	−0.08		3.354
$Ni_{45}Co_5Mn_{30}Ga_{20}$ $(c/a = 1.28)$ "ferri"	3.459	−3.614	0.263	0.619	−0.079		3.268
$Ni_{41}Co_9Mn_{32}Ga_{16}In_2$ $(c/a = 1)$ "ferro"	3.476	3.602	0.405	1.114	−0.098	−0.086	5.185
$Ni_{41}Co_9Mn_{32}Ga_{16}In_2$ $(c/a = 1.28)$ "ferri"	3.418	−3.63	0.261	0.62	−0.083	−0.069	3.279

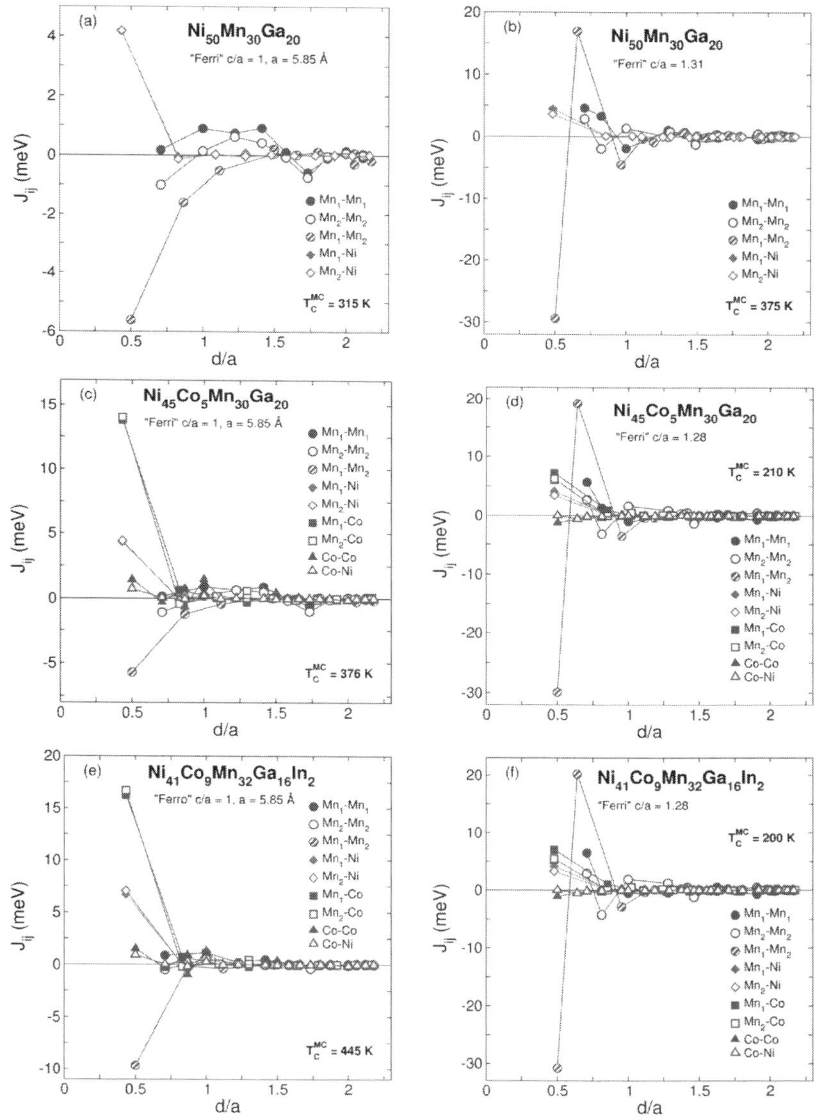

Figure 2. *Ab initio* magnetic exchange interactions of $Ni_{50}Mn_{30}Ga_{20}$ for (**a**) c/a = 1 (cubic), (**b**) c/a = 1.31 (tetragonal), exchange interactions of Co-doped $Ni_{45}Co_5Mn_{30}Ga_{20}$ (**c**) for c/a = 1 and (**d**) c/a = 1.28 and Co- and In-doped $Ni_{41}Co_9Mn_{32}Ga_{16}In_2$ for (**e**) c/a = 1 and (**f**) c/a = 1.28, as a function of the distance d/abetween pairs of atoms i and j (in units of the lattice constant a). TMCC marks the Curie temperatures obtained from MC simulations.

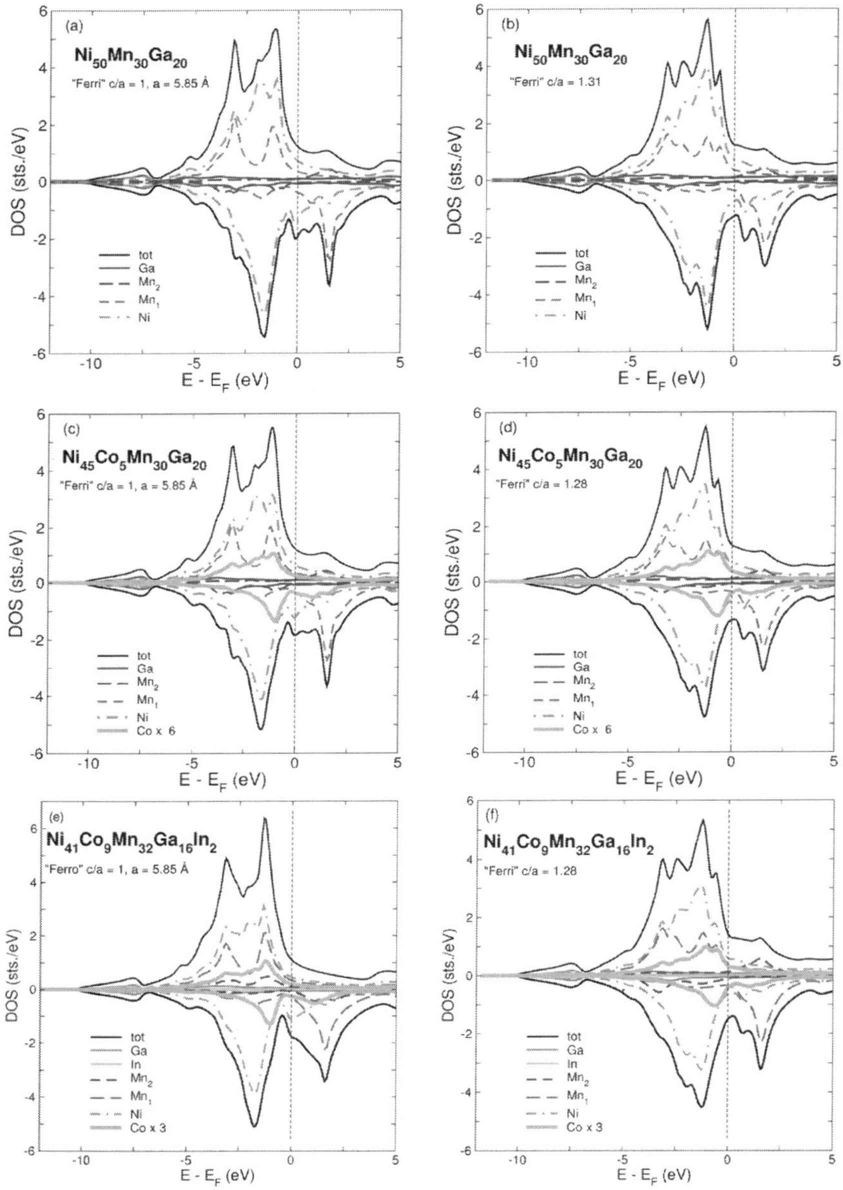

Figure 3. Element resolved density of states of $Ni_{50}Mn_{30}Ga_{20}$ for (**a**) c/a = 1 (cubic), (**b**)c/a = 1.31 (tetragonal), $Ni_{45}Co_5Mn_{30}Ga_{20}$ for (**c**) c/a = 1 and (**d**) c/a = 1.28 and $Ni_{41}Co_9Mn_{32}Ga_{16}In_2$ for (**e**) c/a = 1 and (**f**) c/a = 1.28. "Ferri" means that the spin of Mn on the Ga sites is reversed.

It is obvious from Figure 2 that the exchange coupling constant between neighboring atoms is very sensitive to the inter-atomic distance and drastically changes at crossover from $L2_1$ austenite to $L1_0$ martensite. In general, the strong competition between FM and AFM interactions leads to the oscillatory RKKY (Ruderman–Kittel–Kasuya–Yosida)-type of interactions.

With respect to the J_{ij} interactions of austenitic $Ni_{50}Mn_{30}Ga_{20}$ in Figure 2a, we see that the Mn_1-Mn_1 interaction in the first coordination shell is weaker compared to the following ones. In the first five coordination shells, the Mn_1-Mn_1 interaction is FM, but becomes AFM in the sixth coordination shell. A similar behavior of Mn_1-Mn_1 interactions has been observed for non-stoichiometric $Ni_{2+x}Mn_{1-x}Ga$ alloys [56,62]. The Mn_2-Mn_2 interaction is AFM in the first coordination shell, otherwise it follows the trend of the Mn_1-Mn_1 interaction. For the sixth coordination shell, the Mn_2-Mn_2 interaction becomes also AFM. The Mn_1-Mn_2 interaction is large with AFM interaction up to the forth coordination shell. The origin of large AFM Mn_1-Mn_2coupling is related to the shorter distance between Mn_1 and Mn_2 (d = 0.5a_0) compared to $Mn_{1(2)}$-$Mn_{1(2)}$ (d=2√/2a0). Regarding the $Mn_{1(2)}$-Ni interaction, it is clearly seen that these interactions are strongest in the first coordination shell and then fall off rapidly, which helps to stabilize the FM order. The individual magnetic moments are listed in Table 3. The large value of J_{ij} between Mn is due to the smaller Mn-Ni distance (d=3√/4a0) in contrast to the indirect $Mn_{1(2)}$-$Mn_{1(2)}$coupling.

With respect to the exchange interactions in austenite of Co- and In-doped Ni-Mn-Ga alloys (see Figure 2c,e), the behavior of $Mn_{1(2)}$-$Mn_{1(2)}$, Mn_1-Mn_2 and $Mn_{1(2)}$-Ni interactions is somewhat similar to $L2_1$ $Ni_{50}Mn_{30}Ga_{20}$. The strength of $Mn_{1(2)}$-$Mn_{1(2)}$ interactions slightly decreases, while $Mn_{1(2)}$-Ni coupling constants slightly increase with increasing Co content. For austenite $Ni_{41}Co_9Mn_{32}Ga_{16}In_2$ (see Figure 2e), the nearest neighbor Mn_1-Mn_2 coupling constant is larger by one half compared to $Ni_{50}Mn_{30}Ga_{20}$ and $Ni_{45}Co_5Mn_{30}Ga_{20}$. This is because the J_{ij} calculations have been done for the "ferro" spin configuration due to the fact that the FM reference state is stabilized by increasing Co-content. The strongest $Mn_{1(2)}$-Co interactions are found in austenite of both Co-doped Ni-Mn-Ga alloys, as seen in Figure 2c,e. This is also due to the short distance between Mn and Co atoms (in the case that Co substitutes for Ni).

Regarding the exchange coupling constants of martensite $Ni_{50-y}Co_yMn_{30}Ga_{20}$ (y = 0 and 5 at%) and $Ni_{41}Co_9Mn_{32}Ga_{16}In_2$, which are shown in Figure 2b,d,f, the most striking oscillatory behavior from the FM to AFM interaction is connected with the Mn_1-Mn_2 and $Mn_{1(2)}$-$Mn_{1(2)}$ interactions. For instance, for all alloys, the Mn_1-Mn_2 interaction in the first coordination shell has a large AFM value of $J_{ij} \approx -30$ meV, then it changes sign and reaches a large FM value of $J_{ij} \approx 20$ meV in the next coordination shell. According to the $Mn_{1(2)}$-Co interactions within the first coordination shell, $Mn_{1(2)}$-Ni interactions practically do not change compared to the corresponding interactions for the austenite. The reason for the drastic decrease of the strength in $Mn_{1(2)}$-Co interactions is related to the slight increase of the $Mn_{1(2)}$-Co distance due to the martensitic phase transition. A similar behavior of exchange interactions as a function of the distance between atoms in $Ni_{45}Co_5Mn_{37}In_{13}$ has recently been reported by Comtesse *et al.* [60].

With respect to the different behavior of the individual J_{ij}, a detailed discussion has been presented in the literature by Şaşıoğlu *et al.* [76] in the context of the Anderson model [77]. The magnetic behavior of the X_2YZ Heusler alloys is governed by three types of interactions: (1) the magnetic J_{sd} interaction between the localized *d*- and itinerant *s*–like electrons (this includes Coulomb exchange and hybridization); (2) the magnetic J_{dd} exchange interaction resulting from the interaction between localized and itinerant *d*-like electrons; and (3) a superexchange-like interaction through the Z-(*sp*) electrons. The latter contribution is more important to the indirect exchange mediated mechanism. It is generally accepted that the indirect exchange interaction mediated by the (*sp*)-electrons of the Z-element gives rise to FM order, while the competing superexchange interaction mediated by the same electrons leads to AFM coupling of the Mn-spin moments in the Mn-based Heusler alloys [76,78–83] (although, as stated above, the shorter distance between Mn on its original sublattice and Mn on the Ga sublattice may have AFM contributions from direct magnetic exchange interactions). It should be noted that the Anderson (*s* – *d*)-impurity model for the formation of localized magnetic moments in a metal includes a hybridization term of strength $V_{\mathbf{k}}$, allowing one to promote electrons between the *s*- and *d*-states, as well as a local Coulomb term to avoid the double occupancy of the *d*-electron states [77]. Elimination of odd powers of $V_{\mathbf{k}}$ by successive canonical transformations allows one to derive in a systematic fashion conduction-electron-mediated magnetic interactions between the periodically arranged magnetic impurities. In the fourth order of $V_{\mathbf{k}}$ (*i.e.*, eliminating terms of order $V_{\mathbf{k}}$ and V3k), the RKKY interaction arises from intermediate states involving low-energy spin excitations corresponding to electron-hole pair formation with a spin-flip transition and a superexchange-type of interaction induced by virtual high-energy charge excitations, which promote electrons from the localized d-states to states above E_F [76,84,85]. See Ref. [76] and the original literature [84,85] for a more detailed discussion of these terms. In reciprocal space and at zero temperature, the (strength of) interactions have the following form:

$$J_{RKKY} = \sum_{nn',\mathbf{k}} \left\{ \frac{|V_{n\mathbf{k}}|^2 |V_{n'\mathbf{k}'}|^2}{(\epsilon_{n'\mathbf{k}'} - \epsilon_d)^2} \frac{\Theta(E_F - \epsilon_{n\mathbf{k}})\Theta(\epsilon_{n'\mathbf{k}'} - E_F)}{\epsilon_{n'\mathbf{k}'} - \epsilon_{n\mathbf{k}}} + c.c. \right\} \underset{\mathrm{q}=0}{\longrightarrow} \frac{V^4 \rho(E_F)}{E_d^2} \quad (15)$$

where $\mathbf{k}' = \mathbf{k} + \mathbf{q} + \mathbf{G}$ ($\mathbf{G}$ is a reciprocal lattice vector), $n\mathbf{k}$ labels the band and momentum of the conduction electron Bloch state, ϵ_d is the local energy of the d-electron level and V_{nk} is the hybridization matrix element, which couples conduction and localized d-electrons (in the spirit of the Anderson model, $\varepsilon\uparrow d - \varepsilon\downarrow d$ may be assumed to be of the order of the Hubbard U-energy [77]).

The J_{RKKY} interaction is an oscillating function, and its magnitude is mainly determined by the topology of the Fermi surface and may become rather large in the pre-asymptotic region in the case of the Fermi surface nesting; also the DOS atE_F determines its size. In the $q = 0$ limit, the interaction may be simplified to the limiting value given in Equation (15), where$\rho(E_F)$ is the DOS at E_F and E_d is the energy required to promote an electron from the $3d$ level to E_F . V^4 may be considered as an average over the Fermi surface.

In fourth order, V4k, a second term appears, which describes the superexchange interaction given by:

$$J_{Super-Exch} = \sum_{nn',\mathbf{k}} \left\{ \frac{|V_{n\mathbf{k}}|^2 |V_{n'\mathbf{k}'}|^2}{(\epsilon_{n'\mathbf{k}'} - \epsilon_d)^2} \frac{\Theta(\epsilon_{n\mathbf{k}} - E_F)\Theta(\epsilon_{n'\mathbf{k}'} - E_F)}{\epsilon_{n\mathbf{k}} - \epsilon_d} + c.c. \right\} \quad (16)$$

$$\underset{\mathrm{q}=0}{\longrightarrow} \sum_{n\mathbf{k}} \frac{V^4}{(E_F - \epsilon_{n'\mathbf{k}} - E_d)^3} \quad (17)$$

which is more difficult to estimate without explicit numerical work. We would just like to mention that both types of interactions are simultaneously present in the Heusler alloys and contribute to the magnetic interaction. In the case of disordered alloys, we may find nearest neighbor Mn atoms at shorter interatomic distances, because Mn atoms on the regular Mn-sublattice may be surrounded by Mn on Xand Z-sublattices. This allows for a small overlap of Mn-3*d* wave functions, which will influence the magnetic interactions in addition to the contributions from Equations (15)–(17).

In this work, we have calculated the J_{ij} by using the alternative method based on Liechtenstein' approach [71] for small rotations of the spin moments with frozen potentials, which expresses the exchange parameters in terms of KKR Green's functions for disordered alloys using the CPA. It is worth noting that the CPA neglects short-range correlations. In order to consider the short-range correlations in disordered alloys, we can take into account the so-called non-local coherent potential approximation (NLCPA). However, so far, this has not really been applied to the Heusler alloys, *i.e.*, a detailed investigation of differences between the CPA and the NLCPA has not really been performed so far in the literature. It appears that the KKR-NLCPA provides a basis for systematically including environmental effects within an *ab initio*description of disordered alloys. The implementation proposed by Ebert *et al.* [86] is designed for the treatment of magnetically-ordered alloys. Although there are systematic deviations between NLCPA and CPA, the average NLCPA result, for example for the electronic DOS, is close to that obtained using the standard single-site CPA. The scatter of orbital magnetic moments is greater than for the spin moments. For Heusler alloys with their multi-lattices, it would be interesting to apply the NLCPA, which exhibits strong atomic short-range order effects; however, that is beyond the scope of the present paper, and we have only calculated the J_{ij} by using the single-site CPA. However, we would like to point out that in full-Heusler alloys [87], the short-range interactions have a minimal influence on the electronic structure; therefore, the CPA works well. In general, our results for the stoichiometric and cases of Ni_2Mn_{1+x}(Ga, In, Sn, Sb)$_{1-x}$ (see [55,56,58,59,62]) agree well with those obtained by the frozen-magnon approach to Ni_2MnZ (Z = Ga, In, Sn, Sb) [76,78], apart from small differences.

The calculated total and partial DOS for austenitic and martensitic phases of $Ni_{50}Mn_{30}Ga_{20}$ and $Ni_{45}Co_5Mn_{30}Ga_{20}$ alloys are shown in Figure 3. Obviously, substitution of 5 at% Co for Ni in $Ni_{50}Mn_{30}Ga_{20}$ does practically not change the general shape of the austenitic and martensitic DOS. The antibonding parts of the DOS around and above the Fermi energy E_F have mostly contributions from Mn_1 3*d* states, while bonding parts and non-bonding parts have Mn_1 and Ni 3*d* contributions. Since the concentrations from Mn_2 and Co atoms are small, they contribute little to the total DOS. Nevertheless, in all cases, the small peak in the majority anti-bonding states observed above Fermi E_F is caused by contributions from Mn_2 3*d* states. It is also found that the contributions from the Mn_1 and Mn_2 3*d* states to the total DOS are opposite of each other. For the majority DOS of Mn_1, the main two peaks below E_F are occupied, while for the minority DOS, the antibonding peak is aboveE_F . On the other hand, the partial DOS of Mn_2 below E_F and in the majority DOS above E_F clearly have antiparallel alignment of their spin, which lowers the total magnetic moment. With respect to the partial DOS of Ni, we see that the majority and minority spin states are practically symmetrical, which results in the small magnetic moment of Ni. Another interesting point following from Figure 3 is related to the stabilization of the martensitic phase due to the formation of a pseudogap at the Fermi level. Figure 4 shows the change of DOS for the case that Co, In and Sn concentrations have been increased, showing no dramatic effect due to the increase of *s*, *p* electrons.

4.2. Results of Monte Carlo Simulations

We discuss now the MC simulations of $Ni_{50-y}Co_yMn_{30}Ga_{20}$ (y = 0 and 5 at%) and $Ni_{41}Co_9Mn_{32}Ga_{16}In_2$ using the the extended Potts–Blume–Emery–Griffiths Hamiltonian. For $Ni_{50-y}Co_yMn_{30}Ga_{20}$, we used the magnetic exchange coupling constants J_{ij} calculated for the “ferri” magnetic martensitic and austenitic phases, while for the $Ni_{41}Co_9Mn_{32}Ga_{16}In_2$composition, we have used the J_{ij} of “ferri” martensite and and “ferro” austenite.

Figure 5 shows that the magnetization and strain order parameter coincide for various magnetic fields, pointing to the existence of a coupled magnetostructural phase transition. For $Ni_{50}Mn_{30}Ga_{20}$, the magnetostructural phase transition in the vicinity of 370 K characterizes the transition from ferrimagnetic martensite to paramagnetic austenite. According to the MC simulations and suitably chosen model parameters, the Curie temperature of austenite, T(A)C, is below the Curie temperature of martensite, T(M)C. A similar behavior of magnetization of $Ni_{50}Mn_{30}Ga_{20}$ was found experimentally from isofield magnetic measurements by Fabbrici *et al.* [19,41]. On the other hand, for $Ni_{45}Co_5Mn_{30}Ga_{20}$ and $Ni_{41}Co_9Mn_{32}Ga_{16}In_2$alloys, the magnetostructural transition at $T_m \approx$ 340 K goes along with the transition from weakly magnetic martensite or PM martensite to ferromagnetically ordered austenite.

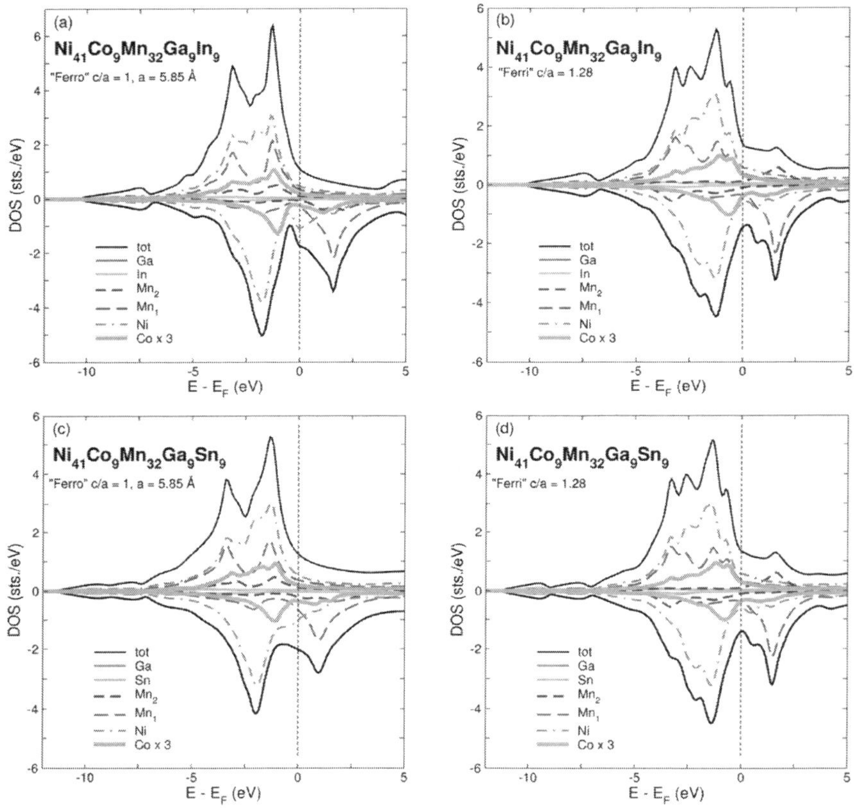

Figure 4. Element resolved density of states of $Ni_{41}Mn_{32}Ga_9In_9$ for (**a**) c/a = 1 (cubic), (**b**) c/a = 1.28 (tetragonal) and $Ni_{41}Co_9Mn_{32}Ga_9Sn_9$ for (**c**) c/a = 1 and (**d**) c/a = 1.28. "Ferri" means that the spin of Mn on the Ga sites is reversed.

With respect to the Curie temperatures of austenite, these are much larger for Co- and In-doped Ni-Mn-Ga alloys compared to the Curie temperatures of martensite. The insets in Figure 5b,c show the isofield magnetization curves for 0.5 and 2 T over a larger temperature range. It is obvious from insets that the T(A)C are equal to ≈ 400 and 445 K for $Ni_{45}Co_5Mn_{30}Ga_{20}$ and $Ni_{41}Co_9Mn_{32}Ga_{16}In_2$ alloys, respectively. Note that these Curie values are close to the experimental data [19,41]. Concerning the Curie temperature of martensite, an increase of Co content results in a decrease of T(M)C. Hence, the paramagnetic gap observed in the vicinity of the structural phase transition is accompanied by an abrupt jump in the magnetization. Note that for $Ni_{50}Mn_{30}Ga_{20}$, the shift of T_m by the magnetic field is positive, while for Co- and In-doped Ni-Mn-Ga alloys, the dT_m/dH_{ext} is negative. Moreover, dT_m/dH_{ext} increases with increasing Co content.

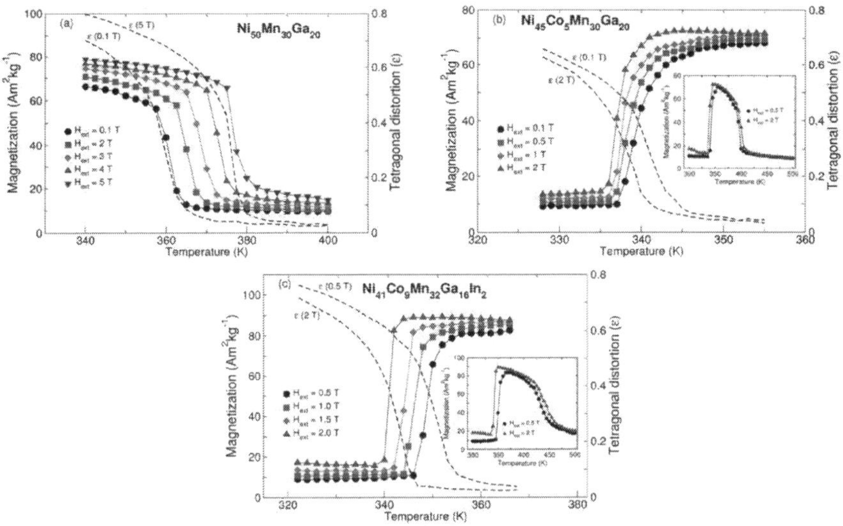

Figure 5. The temperature dependence of calculated magnetization and strain order parameters of (**a**) $Ni_{50}Mn_{30}Ga_{20}$, (**b**) $Ni_{45}Co_5Mn_{30}Ga_{20}$ and (**c**) $Ni_{41}Co_9Mn_{32}Ga_{16}In_2$alloys for various magnetic fields. The insets show the isofield magnetization curves of $Ni_{45}Co_5Mn_{30}Ga_{20}$ and $Ni_{41}Co_9Mn_{32}Ga_{16}In_2$ for 0.5 and 2 T calculated to a greater temperature range.

Concerning the change of magnetization at low temperatures, we would like to note that the Bloch law of spin waves is not obtained for the magnetization in the Monte Carlo simulation [88]. This is a well known result known from spin-fluctuation theories, where the magnetization at low temperatures decays in a linear manner with temperature. However, from the diagonalization of the matrix of the *ab initio* exchange coupling constants, we may obtain, in a mean-field-like fashion, the excitation spectrum (magnon spectrum). This has been done for stoichiometric Ni_2MnGa, but not systematically, in order to incorporate the influence of chemical disorder. Therefore, this has not been included in the paper. Therefore, the isofield-magnetization curves have not been corrected by retaining collective magnetic excitations.

In Figure 6, we present the theoretical temperature dependence of the total specific heat of $Ni_{50-y}Co_yMn_{30}Ga_{20}$ (y = 0 and 5 at%) and $Ni_{41}Co_9Mn_{32}Ga_{16}In_2$ in various magnetic fields. The variation of the specific heat curves is shown in the vicinity of the magnetostructural phase transition. These figures show clearly that in the case of $Ni_{50}Mn_{30}Ga_{20}$, ($Ni_{45}Co_5Mn_{30}Ga_{20}$and $Ni_{41}Co_9Mn_{32}Ga_{16}In_2$), the peaks of specific heat curves are shifted towards higher (lower) temperature when the magnetic field is applied, respectively. For Co- and In-doped alloys, this is a feature of the metamagnetic phase transition from FM to an AFM or PM state upon cooling, which is associated with the magnetostructural phase transition from ordered ferromagnetic austenite to weakly ordered or PM martensite. A comparison of the computational results for the specific heat available from experiments leads to the following observation. The experimental maximal values of specific heat at the T_m of $Ni_{50}Mn_{30}Ga_{20}$, $Ni_{45}Co_5Mn_{30}Ga_{20}$ and $Ni_{41}Co_9Mn_{32}Ga_{16}In_2$ alloys, which are taken for a zero field, are approximately close to 1,750 J/(kg K), 1,600 J/(kg K) and 750 J/(kg K), respectively [19]. Our theoretical values for peaks of specific heat for $Ni_{50}Mn_{30}Ga_{20}$, $Ni_{45}Co_5Mn_{30}Ga_{20}$ and $Ni_{41}Co_9Mn_{32}Ga_{16}In_2$ are 1,720 J/(kg K), 1,640 J/(kg K) and 1,610 J/(kg K), respectively. In general, the theoretical specific heat curves are in qualitative agreement with the experimental data. The slight difference in specific heat values for the In-doped sample can be explained by the abrupt change in the theoretical magnetization across the martensitic transformation in comparison with experimental magnetization behavior, which was more smoothly changed at T_m [19].

The results for the magnetic entropy of $Ni_{50}Mn_{30}Ga_{20}$, $Ni_{45}Co_5Mn_{30}Ga_{20}$, and $Ni_{41}Co_9Mn_{32}Ga_{16}In_2$ alloys in magnetic fields of 0 and 2 T are shown in Figure 7. The curves have been obtained by integrating corresponding specific heat curves using Equation (12). We notice that for $Ni_{50}Mn_{30}Ga_{20}$, the magnetic part of entropy decreases in a magnetic field of 2 T compared to the entropy in a zero field at the same temperature. Besides, the entropy curve is shifted towards a higher temperature by the magnetic field. This feature is typical for magnetic materials showing an FM-PM transition. In contrast, for Co- and In-doped alloys, the entropies in a magnetic field are larger than the entropies in the zero field in the vicinity of the magnetostructural phase transition. Moreover, a magnetic field shifts the entropy curve towards a lower temperature. This points to a metamagnetic phase transition.

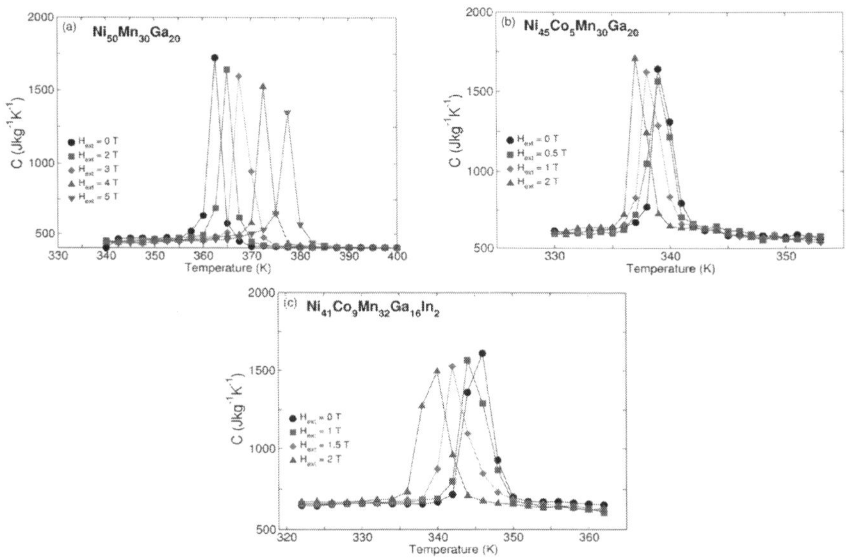

Figure 6. The temperature dependence of total specific heat in the vicinity of the structural phase transformation of (**a**) $Ni_{50}Mn_{30}Ga_{20}$, (**b**) $Ni_{45}Co_5Mn_{30}Ga_{20}$ and (**c**) $Ni_{41}Co_9Mn_{32}Ga_{16}In_2$ alloys for various magnetic fields.

The final computational results for the MCE, isothermal magnetic entropy change ΔS_{mag} and adiabatic temperature change ΔT_{ad} as a function of temperature and magnetic field change from 0 to 2 T are shown in Figure 8 for $Ni_{45}Co_5Mn_{30}Ga_{20}$ and $Ni_{41}Co_9Mn_{32}Ga_{16}In_2$. In the case of $Ni_{50}Mn_{30}Ga_{20}$, the value of ΔS_{mag} is negative, while the value of ΔT_{ad} is positive at the transition temperature. This is a feature of the conventional MCE related to the coupled magnetostructural phase transition from magnetically ordered martensite to magnetically disordered austenite. In contrast, for Co- and In-doped alloys, we observe the inverse MCE characterized by $\Delta T_{ad} < 0$ and $\Delta S_{mag} > 0$. In order to compare the theoretical values of MCE with experimental ones [19], we have also inserted the experimental MCE curves by dashed lines with open symbols in Figure 8. It is obvious from Figure 8 that there is a qualitative agreement between the theoretical results obtained in the framework of *ab initio* calculations and MC simulations using the Potts-BEG model and the experimental data. The slight difference in MCE values between theoretical and experimental results can be attributed to the abrupt change in the theoretical magnetization curves across the martensitic transformation compared with the 'softer' experimental magnetization curves, which change smoothly across T_m [19,48].

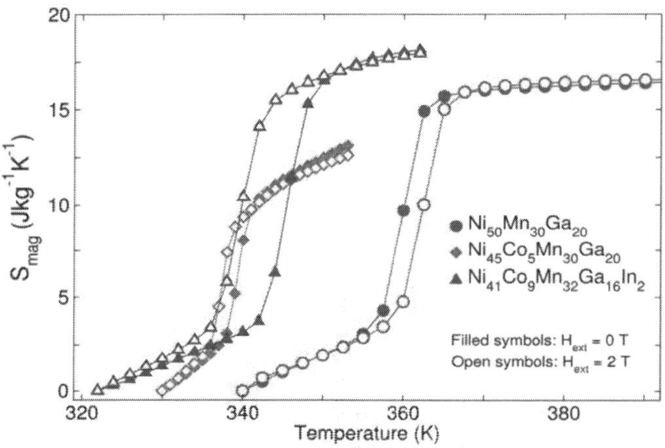

Figure 7. The temperature dependence of the isothermal magnetic entropy calculated for $Ni_{50-y}Co_yMn_{30}Ga_{20}$ (y = 0 and 5 at%) and $Ni_{41}Co_9Mn_{32}Ga_{16}In_2$ alloys in magnetic fields of 0 and 2 T.

5. CONCLUSIONS

In this paper, we presented a theoretical study of the magnetic and magnetocaloric properties of Co- and In-doped Ni-Mn-Ga alloys by combining *ab initio* calculations and Monte Carlo simulations. In the first step, we have done the first-principles calculations of magnetic exchange coupling constants and DOS of $Ni_{50-y}Co_yMn_{30}Ga_{20}$ (y = 0 and 5 at%) and $Ni_{41}Co_9Mn_{32}Ga_{16}In_2$ alloys for different structural and magnetic configurations. Our previous results have shown that ferrimagnetic order is favorable for both austenitic and martensitic phases in $Ni_{50-y}Co_yMn_{30}Ga_{20}$, while for $Ni_{41}Co_9Mn_{32}Ga_{16}In_2$, the ferromagnetic reference state in austenite and the ferrimagnetic reference state in martensite are favorable. For all alloys, we have found the strong competition of FM and AFM exchange interactions in the martensitic phase. The addition of Co leads to very large FM interactions between Mn and Co in the austenite phase, whereas for the martensite phase, these interactions are smaller by half those for austenite. In the second step, in order to obtain the finite temperature dependence of the magnetic and magnetocaloric properties of Ni-(Co)-Mn-Ga-(In) alloys, we have used the values of exchange couplings and magnetic moments as input parameters for the microscopic Potts-BEG Hamiltonian. As mentioned before, the addition of Co to Ni-Mn-Ga alloys results in the splitting of coupled phase transitions at the Curie temperature. As a result, the paramagnetic gap in the vicinity of the martensitic transformation temperature and the metamagnetic behavior of magnetization across the martensitic transformation are observed. Concerning the Curie temperatures, TAC is found to increase, whereas TMC decreased with increasing Co content. This behavior can be attributed to the stronger Co-Mn, Mn-Mn and Mn-Ni exchange couplings in austenite compared to the weaker ones in martensite. Moreover, the crossover from a direct to inverse magnetocaloric effect in Ni-Mn-Ga is achieved due to the substitution of Ni by Co, resulting in the appearance of a paramagnetic gap in martensite. Simulation of the magnetic and magnetocaloric properties of Co-doped Ni-Mn-Ga are in a good qualitative agreement with the available experimental data.

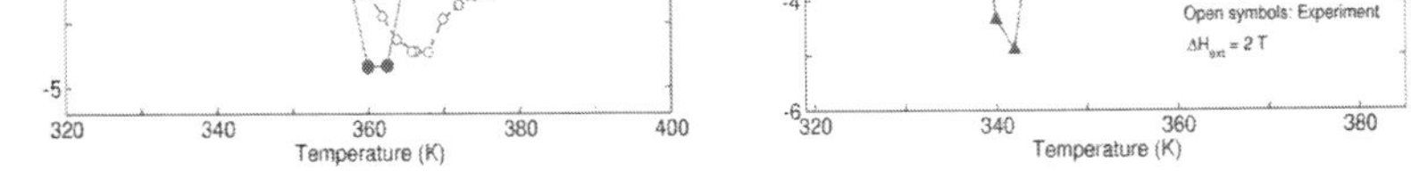

Figure 8. The temperature dependence of (**a**) isothermal magnetic entropy change and (**b**) adiabatic temperature change for $Ni_{50-y}Co_yMn_{30}Ga_{20}$ (y = 0 and 5 at%) and $Ni_{41}Co_9Mn_{32}Ga_{16}In_2$ alloys upon variation of the magnetic field from 0 to 2 T. Experimental magnetocaloric curves are shown by dashed lines with open symbols and were measured in a magnetic field change ΔH_{ext} of 1.8 T. Experimental data for $Ni_{50-y}Co_yMn_{30}Ga_{20}$ (y = 0 and 5 at%) have been taken from [19], while for the $Ni_{41}Co_9Mn_{32}Ga_{16}In_2$ alloy, the data have been taken from [48].

ACKNOWLEDGMENTS

This work is supported by Russian Foundation Basic Research (RFBR) Grant No. 14-02-01085, Russian Science Foundation (RSF) No. 14-12-00570\14 (Section 2), Ministry of Education and Science of Russian Federation (RF) No 3.2021.2014/K (Sections 3 and 4) and the Creation and Development Program of National University of Science and Technology (NUST) "MISiS". Peter Entel and Anna Grünebohm acknowledge Deutsche Forschungsgemeinschaft (DFG) (Schwerpunktsprogramm (SPP) 1599) for financial support.

AUTHOR CONTRIBUTIONS

Vladimir Sokolovkiy and Peter Entel wrote the manuscript. Vasiliy Buchelnikov, Anna Grünebohm, and Vladimir Sokolovskiy performed the *ab initio* calculations. Vladimir Sokolovskiy performed the Monte Carlo simulations. All authors have read and approved the final manuscript.

REFERENCES

1. Vasiliev, A.N.; Buchelnikov, V.D.; Takagi, T.; Khovailo, V.V.; Estrin, E.I. Shape-memory ferromagnets. *Phys.-Uspekhi***2003**, *46*, 559–588

2. Planes, A.; Mañosa, L. Ferromagnetic shape-memory alloys. *Mater. Sci. Forum* **2006**, *512*, 145–152.

3. Buchelnikov, V.D.; Vasiliev, A.N.; Koledov, V.V.; Taskaev, S.V.; Khovaylo, V.V.; Shavrov, V.G. Magnetic shape-memory alloys: Phase transitions and functional properties. *Phys.-Uspekhi* **2006**, *49*, 871–877.

4. Entel, P.; Buchelnikov, V.D.; Gruner, M.E.; Hucht, A.; Khovailo, V.V.; Nayak, S.K.; Zayak, A.T. Shape Memory Alloys: A Summary of Recent Achievements. *Mater. Sci. Forum* **2008**, *583*, 21–41.

5. Planes, A.; Mañosa, L.; Acet, M. Magnetocaloric effect and its relation to shape-memory properties in ferromagnetic Heusler alloys. *J. Phys. Condens. Matter* **2009**, *21*.

6. Martynov, V.V.; Kokorin, V.V. The crystal structure of thermally- and stress-induced martensites in Ni_2MnGa single crystals. *J. Phys. III* **1992**, *2*, 739–749.

7. Chernenko, V.A.; Amengual, A.; Cesari, E.; Kokorin, V.V.; Zasimchuk, I.K. Thermal and magnetic properties of stress-induced martensites in Ni-Mn-Ga alloys. *J. Phys. IV* **1995**, *5*.

8. Ullakko, K.; Huang, J.K.; Kantner, C.; O'Handley, R.C. Large magnetic-field-induced strains in Ni_2MnGa single crystals. *Appl. Phys. Lett* **1996**, *10*, 1966–1968.

9. Murray, S.J.; Marioni, M.A.; Allen, S.M.; O'Handley, R.C.; Lograsso, T.A. 6% magnetic-field-induced strain by twin-boundary motion in ferromagnetic Ni-Mn-Ga. *Appl. Phys. Lett* **2000**, *77*, 886–888.

10. Sozinov, A.; Likhachev, A.; Lanska, N.; Ullakko, K. Giant magnetic-field-induced strain in NiMnGa seven-layered martensitic phase. *Appl. Phys. Lett* **2002**, *80*, 1746–1748.

11. Sozinov, A.; Likhachev, A.A.; Lanska, N.; Ullakko, K.; Lindroos, V.K. 10% magnetic-field-induced strain in Ni-Mn-Ga seven-layered martensite. *J. Phys. IV* **2003**, *112*, 955–958.

12. Karaca, H.E.; Karaman, I.; Basaran, B.; Chumlyakov, Y.I.; Maier, H.J. Magnetic field and stress induced martensite reorientation in NiMnGa ferromagnetic shape memory alloy single crystal. *Acta Mater* **2006**, *54*, 233–245.

13. Martynov, V.V.; Kokorin, V.V. Shape memory and multistage superelasticity in Heusler-type Ni-Mn-Ga single crystals. *Trans. Mater. Res. Soc. Jpn* **1994**, *18B*, 839–844.

14. Chernenko, V.A.; L'vov, V.; Pons, J.; Cesari, E. Superelasticity in high-temperature Ni-Mn-Ga alloys. *J. Appl. Phys* **2003**,*93*, 2394–2399.

15. L'vov, V.A.; Rudenko, A.A.; Chernenko, V.A.; Cesari, E.; Pons, J.; Kanomata, T. Stress-induced martensitic transformation and superelasticity of alloys: Experiment and theory. *Mater. Trans* **2006**, *46*, 790–797.

16. Aliev, A.; Batdalov, A.; Bosko, S.; Buchelnikov, V.D.; Dikshtein, I.; Khovailo, V.; Koledov, V.; Levetin, R.; Shavrov, V.; Takagi, T. Magnetocaloric effect and magnetization in Ni-Mn-Ga Heusler alloy in the vicinity of magnetostructural transition. *J. Magn. Magn. Mater* **2004**, *272*, 2040–2042.

17. Albertini, F.; Paoluzi, A.; Pareti, L.; Solzi, M.; Villa, E.; Besseghini, S.; Passaretti, F. Phase transition and magnetocaloric entropy change in Mn-rich Ni_2MnGa alloys. *J. Appl. Phys* **2006**, *100*.

18. Khovaylo, V.V.; Skokov, K.P.; Koshkid'ko, Y.S.; Koledov, V.V.; Shavrov, V.G.; Buchelnikov, V.D.; Taskaev, S.V.; Miki, H.; Takagi, T.; Vasiliev, A.N. Adiabatic temperature change at first-order magnetic phase transitions: $Ni_{2.19}Mn_{0.81}Ga$ as a case study. *Phys. Rev. B* **2008**, *78*.

19. Fabbrici, S.; Porcari, G.; Cugini, F.; Solzi, M.; Kamarad, J.; Arnold, Z.; Cabassi, R.; Albertini, F. Co and In doped Ni-Mn-Ga magnetic shape memory alloys: A thorough structural, magnetic and magnetocaloric study. *Entropy* **2014**, *16*, 2204–2222.

20. Han, Z.D.; Qian, B.; Wang, D.H.; Zhang, P.; Jiang, X.F.; Zhang, C.L.; Du, Y.W. Magnetic phase separation and exchange bias in off-stoichiometric Ni-Mn-Ga alloys. *Appl. Phys. Lett* **2013**, *103*.

21. Çakr, A.; Righi, L.; Albertini, F.; Acet, M.; Farle, M.; Aktúrk, S. Extended investigation of intermartensitic transitions in Ni-Mn-Ga magnetic shape memory alloys: A detailed phase diagram determination. *J. Appl. Phys* **2013**, *114*.

22. Wang, Y.; Huang, C.; Gao, J.; Yang, S.; Ding, X.; Song, X.; Ren, X. Evidence for ferromagnetic strain glass in Ni-Co-Mn-Ga Heusler alloy system. *Appl. Phys. Lett* **2012**, *101*.

23. Wang, Y.; Huang, C.; Wu, H.; Gao, J.; Yang, S.; Wang, D.; Ding, X.; Song, X.; Ren, X. Spontaneous strain glass to martensite transition in ferromagnetic Ni-Co-Mn-Ga strain glass. *Appl. Phys. Lett* **2013**, *102*.

24. Bozhko, A.D.; Vasiliev, A.N.; Khovailo, V.V.; Buchelnikov, V.D.; Dikshtein, I.E.; Seletskii, S.M.; Shavrov, V.G. Phase transitions in the ferromagnetic alloys Ni_{2+x} Mn_{1-x} Ga. *JETP Lett* **1998**, *67*, 227–232.

25. Jin, X.; Marioni, M.; Bono, D.; Allen, S.M.; O'Handley, R.C.; Hsu, T.Y. Empirical mapping of Ni-Mn-Ga properties with composition and valence electron concentration. *J. Appl. Phys* **2002**, *91*, 8222–8224.

26. Khovailo, V.V.; Novosad, V.; Takagi, T.; Filippov, D.A.; Levetin, R.Z.; Vasil'ev, A.N. Magnetic properties and magnetostructural phase transitions in Ni_{2+x} Mn_{1-x} Ga shape memory alloys. *Phys. Rev. B* **2004**, *70*.

27. Khovaylo, V.V.; Buchelnikov, V.D.; Kainuma, R.; Koledov, V.V.; Otsuka, M.; Shavrov, V.G.; Takagi, T.; Taskaev, S.V.; Vasiliev, A.N. Phase transitions in Ni_{2+x} Mn_{1-x} Ga with a high Ni excess. *Phys. Rev. B* **2005**, *72*.

28. Richard, M.; Feuchtwanger, J.; Schlagel, D.; Lograsso, T.; Allen, S.M.; O'Handley, R.C. Crystal structure and transformation behavior of Ni-Mn-Ga martensites. *Scr. Mater* **2006**, *54*, 1797–1801.

29. Li, Z.; Zhang, Y.; Esling, C.; Zhao, X.; Zuo, L. Twin relationships of 5M modulated martensite in Ni-Mn-Ga alloy. *Acta Mater* **2011**, *59*, 3390–3397.

30. Ranjan, R.; Banik, S.; Barman, S.R.; Kumar, U.; Mukhopadhyay, P.K.; Pandey, D. Powder x-ray diffraction study of the thermoelastic martensitic transition on $Ni_2Mn_{1.05}Ga_{0.95}$. *Phys. Rev. B* **2006**, *74*.

31. Righi, L.; Albertini, F.; Villa, E.; Paoluzi, A.; Calestani, G.; Chernenko, V.; Besseghini, S.; Ritter, C.; Passaretti, F. Crystal structure of 7M modulated Ni-Mn-Ga martensitic phase. *Acta Mater* **2008**, *56*, 4529–4535.

32. Wedel, B.; Suzuki, M.; Murakami, Y.; Wedel, C.; Suzuki, T.; Shindo, D.; Itagaki, K. Low temperature crystal structure of Ni-Mn-Ga alloys. *J. Alloys Compd* **1999**, *290*, 137–143.

33. Söderberg, O.; Brown, D.; Aaltio, I.; Oksanen, J.; Syrén, J.; Pulkkinen, H.; Hannula, S.P. Microstructure and properties of Ni-Mn-Ga alloys produced by rapid solidification and pulsed electric current sintering. *J. Alloys Compd* **2011**, *509*, 5981–5987.

34. Khovailo, V.V.; Abe, T.; Koledov, V.V.; Matsumoto, M.; Nakamura, H.; Note, R.; Ohtsuka, M.; Shavrov, V.G.; Takagi, T. Influence of Fe and Co on Phase Transitions in Ni-Mn-Ga Alloys. *Mater. Trans* **2003**, *44*, 2509–2512.

35. Cong, D.Y.; Wang, S.; Wang, Y.D.; Ren, Y.; Zuo, L.; Esling, C. Martensitic and magnetic transformation in Ni-Mn-Ga-Co

ferromagnetic shape memory alloys. *Mater. Sci. Eng. A* **2008**, *473*, 213–218.

36. Ma, Y.; Yang, S.; Liu, Y.; Liu, X. The ductility and shape-memory properties of Ni-Mn-Co-Ga high-temperature shape-memory alloys. *Acta Mater* **2009**, *57*, 3232–3241.

37. Glavatskyy, I.; Glavatska, N.; Söderberg, O.; Hannula, S.-P.; Hoffmann, J.-U. Transformation temperatures and magnetoplasticity of Ni-Mn-Ga alloyed with Si, In, Co or Fe. *Scr. Mater* **2006**, *54*, 1891–1895.

38. Soto-Parra, D.E.; Moya, X.; Mañosa, L.; Planes, A.; Flores-Zúñiga, H.; Alvarado-Hernández, F.; Ochoa-Gamboa, R.A.; Matutes-Aquino, J.A.; Ríos-Jara, D. Fe and Co selective substitution in Ni_2MnGa: Effect of magnetism on relative phase stability. *Philos. Mag* **2010**, *90*, 2771–2792.

39. Rolfs, K.; Wimpory, R.C.; Petry, W.; Schneider, R. Effect of alloying Ni-Mn-Ga with Cobalt on thermal and structural properties. *J. Phys. Conf. Ser* **2010**, *251*.

40. Kumar, A.S.; Ramudu, M.; Seshubai, V. Effect of selective substitution of Co for Ni or Mn on the superstructure and microstructural properties of $Ni_{50}Mn_{29}Ga_{21}$. *J. Alloys Compd* **2011**, *509*, 8215–8222

41. Fabbrici, S.; Kamarad, J.; Arnold, Z.; Casoli, F.; Paoluzi, A.; Bolzoni, F.; Cabassi, R.; Solzi, M.; Porcari, G.; Pernechele, C.; *et al.* From direct to inverse giant magnetocaloric effect in Co-doped NiMnGa multifunctional alloys. *Acta Mater***2011**, *59*, 412–419.

42. Kumar, A.S.; Ramudu, M.; Seshubai, V. Structural and magnetic investigations in the vicinity of first-order transformations in Ni-Mn-Ga-Co ferromagnetic shape memory alloys. *Phase Trans* **2012**, *85*, 1045–1059.

43. Segui, C.; Cesari, E. Composition and atomic order effects on the structural and magnetic transformations in ferromagnetic Ni-Co-Mn-Ga shape memory alloys. *J. Appl. Phys* **2012**, *111*.

44. Pushin, V.G.; Korolev, A.V.; Kourov, N.I.; Kuranova, N.N.; Marchenkova, E.B.; Popov, A.G. On the effect of cobalt doping on thermoelastic martensitic transformations in ferromagnetic Heusler Ni_{50-x} Co_x $Mn_{29}Ga_{21}$ magnetically controlled shape memory alloys. *Tech. Phys. Lett* **2013**, *39*, 737–740.

45. Khan, M.; Dubenko, I.; Stadler, S.; Ali, N. The structural and magnetic properties of $Ni_2Mn_{1-x}M_xGa$ (M = Co, Cu). *J. Appl. Phys* **2005**, *97*.

46. Kanomata, T.; Nunoki, S.; Endo, K.; Kataoka, M.; Nishihara, H.; Khovaylo, V.V.; Umetsu, R.Y.; Shishido, T.; Nagasako, M.; Kainuma, R.; *et al.* Phase diagram of the ferromagnetic shape memory alloys $Ni_2MnGa_{1x}Co_x$. *Phys. Rev. B* **2012**,*85*.

47. Kumar, A.S.; Ramudu, M.; Seshubai, V. Site preference of magnetic atoms in Ni-Mn-Ga-M (M = Co, Fe) ferromagnetic shape memory alloys. *Phys. Status Solidi B* **2012**, *249*, 620–626.

48. Porcari, G.; Fabbrici, S.; Pernechele, C.; Albertini, F.; Buzzi, M.; Paoluzi, A.; Kamarad, J.; Arnold, Z.; Solzi, M. Reverse magnetostructural transformation and adiabatic temperature change in Coand In-substituted Ni-Mn-Ga alloys. *Phys. Rev. B* **2012**, *85*.

49. Emre, B.; Yüce, S.; Stern-Taulats, E.; Planes, A.; Fabbrici, S.; Albertini, F.; Mañosa, L. Large reversible entropy change at the inverse magnetocaloric effect in Ni-Co-Mn-Ga-In magnetic shape memory alloys. *J. Appl. Phys* **2013**, *113*.

50. Segui, C. Effects of the interplay between atomic and magnetic order on the properties of metamagnetic Ni-Co-Mn-Ga shape memory alloys. *J. Appl. Phys* **2014**, *115*.

51. Entel, P.; Gruner, M.E.; Adeagbo, W.A.; Zayak, A.T. Magnetic field-induced changes in magnetic shape memory alloys. *Mater. Sci. Eng. A* **2008**, *481–482*, 258–261.

52. Tan, C.-L.; Jiang, J.-X.; Tian, X.-H.; Cai, W. Effect of Co on magnetic property and phase stability of Ni-Mn-Ga ferromagnetic shape-memory alloys: A first-principles study. *Chin. Phys. B* **2010**, *19*.

53. Bai, J.; Raulot, J.-M.; Zhang, Y.; Esling, C.; Zhao, X.; Zuo, L. The effects of alloying element Co on Ni-Mn-Ga ferromagnetic shape memory alloys from first-principles calculations. *Appl. Phys. Lett* **2011**, *98*.

54. Zelený, M.; Sozinov, A.; Straka, L.; Björkman, T.; Nieminen, R.M. First-principles study of Co- and Cu-doped Ni_2MnGa along the tetragonal deformation path. *Phys. Rev. B* **2014**, *89*.

55. Buchelnikov, V.D.; Entel, P.; Taskaev, S.V.; Sokolovskiy, V.V.; Hucht, A.; Ogura, M.; Akai, H.; Gruner, M.E.; Nayak, S.K. Monte Carlo study of the influence of antiferromagnetic exchange interactions on the phase transitions of ferromagnetic Ni-Mn-X alloys (X = In, Sn, Sb). *Phys. Rev. B* **2008**, *78*.

56. Entel, P.; Dannenberg, A.; Siewert, M.; Herper, H.C.; Gruner, M.E.; Buchelnikov, V.D.; Chernenko, V.V. Composition-dependent basics of smart Heusler materials from first-principles calculations. *Mater. Sci. Forum* **2011**, *684*, 1–29.

57. Li, C.-M.; Luo, H.-B.; Hu, Q.-M.; Yang, R.; Johansson, B.; Vitos, L. Role of magnetic and atomic ordering in the martensitic transformation of Ni-Mn-In from a first-principles study. *Phys. Rev. B* **2012**, *86*.

58. Sokolovskiy, V.V.; Buchelnikov, V.D.; Zagrebin, M.A.; Entel, P.; Sahool, S.; Ogura, M. First-principles investigation of chemical and structural disorder in magnetic $Ni_2Mn_{1+x}Sn_{1-x}$ Heusler alloys. *Phys. Rev. B* **2012**, *86*.

59. Entel, P.; Siewert, M.; Gruner, M.E.; Herper, H.C.; Comtesse, D.; Arróyave, R.; Singh, N.; Talaparta, A.; Sokolovskiy, V.V.; Buchelnikov, V.D.; *et al.* Complex magnetic ordering as a driven mechanism of multifunctional properties of Heusler alloys from first-principles. *Eur. Phys. J. B* **2013**, *86*, 65–75.

60. Comtesse, D.; Gruner, M.E.; Ogura, M.; Sokolovskiy, V.V.; Buchelnikov, V.D.; Grünebohm, A.; Arróyave, R.; Singh, N.; Gottschall, T.; Gutfleisch, O.; *et al.* First-principles calculation of the instability leading to giant inverse magnetocaloric effects. *Phys. Rev. B* **2014**, *89*.

61. Castán, T.; Vives, E.; Lindgard, P.-A. Modeling premartensitic effects in Ni_2MnGa: A mean-field and Monte Carlo simulation study. *Phys. Rev. B* **1999**, *60*.

62. Buchelnikov, V.D.; Sokolovskiy, V.V.; Herper, H.C.; Ebert, H.; Gruner, M.E.; Taskaev, S.V.; Khovaylo, V.V.; Hucht, A.; Dannenberg, A.; Ogura, M.; *et al.* First-principles and Monte Carlo study of magnetostructural transition and magnetocaloric properties of $Ni_{2+x}Mn_{1-x}Ga$. *Phys. Rev. B* **2010**, *81*.

63. Buchelnikov, V.D.; Sokolovskiy, V.V.; Taskaev, S.V.; Khovaylo, V.V.; Aliev, A.A.; Khanov, L.N.; Batdalov, A.B.; Entel, P.; Miki, H.; Takagi, T.

Monte Carlo simulations of the magnetocaloric effect in magnetic Ni-Mn-X (X = Ga, In) Heusler alloys. *J. Phys. D Appl. Phys* **2011**, *44*.

64. Sokolovskiy, V.V.; Buchelnikov, V.D.; Taskaev, S.V.; Khovaylo, V.V.; Ogura, M.; Entel, P. Quaternary Ni-Mn-In-Y Heusler alloys: A way to achieve materials with better magnetocaloric properties? *J. Phys. D Appl. Phys* **2013**, *46*.

65. Singh, N.; Arróyave, R. Magnetocaloric effects in Ni-Mn-Ga-Fe alloys using Monte Carlo simulations. *J. Appl. Phys***2013**, *113*.

66. Entel, P.; Gruner, M.; Comtesse, D.; Sokolovskiy, V.V,; Buchelnikov, V.D. Interacting magnetic cluster-spin glasses and strain glasses in Ni-Mn based Heusler structured intermetallics. *Phys. Status Solidi* **2014**, *1–14*.

67. Ray, D.K.; Jardin, J.P. Elastic and magnetic interactions in a narrow twofold-degenerate band. *Phys. Rev. B* **1986**, *33*, 5021–5027.

68. Murakami, Y.; Kainuma, R.; Shindo, D.; Tonomura, A. Magnetic microstructure analysis of ferromagnetic shape memory alloys and related compounds. *Mater. Sci. Forum* **2010**, *684*, 117–128.

69. Ebert, H.; Ködderitzsch, D.; Minár, J. Calculating condensed matter properties using the KKR-Green's function method-recent developments and applications. *Rep. Prog. Phys* **2011**, *74*.

70. Ebert, H. SPR-KKR Package, *version 6.3; allowing to calculate the electronic structure of various periodic systems, including in particular systems with chemical disorder,* Available online: http://ebert.cup.unimuenchen.de accessed on 17 September 2014.

71. Liechtenstein, A.I.; Katsnelson, M.I.; Antropov, V.P.; Gubanov, V.A. Local spin-density functional-approach to the theory of exchange interactions in ferromagnetic metals and alloys. *J. Magn. Magn. Mater* **1987**, *67*, 65–74.

72. Perdew, J.P.; Burke, K.; Enzerhof, M. Generalized gradient approximation made simple. *Phys. Rev. Lett* **1996**, *77*, 3865–3868.

73. Albertini, F.; Pareti, L.; Paoluzi, A.; Morellon, L.; Algaarabel, P.A.; Ibarra, M.R.; Righi, L. Composition and temperatue dependence of the magnetocrystalline anisotropy in Ni_{2+x} Mn_{1-y} Ga_{1-z} $(x + y + z = 0)$ Heusler alloys. *Appl. Phys. Lett***2002**, *81*, 4032–4034.

74. Enkovaara, J.; Ayuela, A.; Nordstöm, L.; Nieminen, R.M. Magnetic anisotropy in Ni_2MnGa. *Phys. Rev. B* **2002**, *65*.

75. Gruner, M.E.; Entel, P.; Opahle, I.; Richter, M. Ab initio investigation of twin boundary motion in the magnetic shape memory Heusler alloy Ni_2MnGa. *J. Mater. Sci* **2008**, *43*, 3825–3831.

76. Şaşıoğlu, E.; Sandratskii, L.M.; Bruno, P. Role of conduction electrons in mediating exchange interactions in Mn-based Heusler alloys. *Phys. Rev. B* **2008**, *77*.

77. Anderson, P.W. Localized magnetic states in metals. *Phys. Rev. B* **1961**, *124*.

78. Şaşıoğlu, E.; Sandratskii, L.M.; Bruno,, P. First-principles calculation of the intersublattice exchange interactions and Curie temperatures of the full Heusler alloys Ni_2MnX (X = Ga, In, Sn, Sb). *Phys. Rev. B* **2004**, *70*.

79. Kurtulus, Y.; Dronskowski, R.; Samolyuk, G.D.; Antropov, V.P. Electronic structure and magnetic exchange coupling in ferromagnetic full Heusler alloys. *Phys. Rev. B* **2005**, *71*.

80. Rusz, J.; Bergqvist, L.; Kudrnovský, J.; Turek, I. Exchange interactions and Curie temperatures in Ni_{2-x} MnSb alloys: First-principles study. *Phys. Rev. B* **2006**, *73*.

81. Şaşıoğlu, E. Nonzero macroscopic magnetization in half-metallic antiferromagnets at finite temperatures. *Phys. Rev. B***2009**, *79*.

82. Meinert, M.; Schmalhorst, J.-M.; Reiss, G. *Ab initio* prediction of ferrimagnetism, exchange interactions and Curie temperatures in Mn_2TiZ Heusler compounds. *J. Phys. Condens. Matter* **2011**, *23*.

83. Meinert, M.; Schmalhorst, J.-M.; Reiss, G. Exchange interactions and Curie temperatures of Mn_2CoZ compounds. *J. Phys. Condens. Matter* **2011**, *23*.

84. Da Silva, C.G.; Falicov, L.M. Theory of magnetic properties of rare earth compounds (Localized moments and hybridization effects). *J. Phys. C* **1972**, *5*, 63–76.

85. Shi, Z.-P.; Levy, P.M.; Fry, J.L. Interlayer magnetic coupling in metallic multilayer structures. *Phys. Rev. B* **1994**, *49*.

86. Ködderitzsch, D.; Ebert, H.; Rowlands, D.A.; Ernst, A. Relativistic formulation of the Korringa-Kohn-Rostoker nonlocal coherent-potential approximation. *New J. Phys* **2007**, *9*, 81–17.

87. Özdŏgan, K.; Şaşıoğlu, E.; Galanakis, I. Engineering the electronic, magnetic, and gap-related properties of the quinternary half-metallic Heusler alloys. *J. Appl. Phys* **2008**, *103*.

88. Galanakis, I.; Şaşıoğlu, E. Variation of the magnetic properties of Ni_2MnGa Heusler alloy upon tetragonalization: a first-principles study. *J. Phys. D Appl. Phys* **2011**, *44*.

8

PHASE COMPETITIONS BEHIND THE GIANT MAGNETIC ENTROPY VARIATION: $GD_5SI_2GE_2$ AND $TB_5SI_2GE_2$ CASE STUDIES

Ana Lúcia Pires [1,2,†], João Horta Belo [1,†], Armandina Maria Lima Lopes [1,2], Isabel T. Gomes [1], Luis Morellón [3], Cesar Magen [3,4], Pedro Antonio Algarabel [5], Manuel Ricardo Ibarra [3], André Miguel Pereira [1,*] and João Pedro Araújo [1,*]

[1] IFIMUP and IN—Institute of Nanoscience and Nanotechnology, Departamento de Física e Astronomia da Faculdade de Ciências da Universidade do Porto, Rua do Campo Alegre, 687, Porto 4769-007, Portugal; E-Mials:

[2] CFNUL—Centro de Física Nuclear da Universidade de Lisboa, Av. Prof. Gama Pinto, 2, Lisboa 1649-003, Portugal

[3] Instituto de Nanociencia de Aragón, Universidad de Zaragoza, ES-50018 Zaragoza, Spain and Departamento de Física de la Materia Condensada, Universidad de Zaragoza, Zaragoza 50009, Spain

[4] Fundación ARAID, Zaragoza 50018, Spain

[5] Instituto de Ciencia de Materiales de Aragón, CSIC-Universidad de Zaragoza, and Departamento de Física de la Materia Condensada, Universidad de Zaragoza, Zaragoza 50009, Spain

ABSTRACT

Magnetic materials with strong spin-lattice coupling are a powerful set of candidates for multifunctional applications because of their multiferroic, magnetocaloric (MCE), magnetostrictive and magnetoresistive effects. In these materials there is a strong competition between two states (where a state comprises an atomic and an associated magnetic structure) that leads to the occurrence of phase transitions under subtle variations of external parameters, such as temperature, magnetic field and hydrostatic pressure. In this review a general method combining detailed magnetic measurements/analysis and first principles calculations with the purpose of estimating the phase transition temperature is presented with the help of two examples ($Gd_5Si_2Ge_2$ and $Tb_5Si_2Ge_2$). It is demonstrated that such method is an important tool for a deeper understanding of the (de)coupled nature of each phase transition in the materials belonging to the $R_5(Si,Ge)_4$ family and most possibly can be applied to other systems. The exotic Griffiths-like phase in the framework of the $R_5(Si_xGe_{1-x})_4$ compounds is reviewed and its generalization as a requisite for strong phase competitions systems that present large magneto-responsive properties is proposed.

KEYWORDS

entropy; phase transitions; magnetocaloric effect

1. INTRODUCTION

Back in the 1980s, the simultaneous discovery of the "ozone hole" in the stratosphere and of the influence of the emission of CFC gases, widely used in every refrigeration system at that time, in the thinning of the ozone layer triggered the search for alternative refrigeration technologies in order to replace the gas-compression technology [1]. The thermoelectric and thermomagnetic technologies are ahead in this race. Here only the latter one will be discussed and in particular the nature of the effect behind this technology—the magnetocaloric effect (MCE). The MCE was first observed in the 1840s, when Joule observed a temperature variation in a sample of pure iron as it was subjected to a magnetic field change [2]. It was understood on the basis of thermodynamics by Weiss and Piccard in 1917 [2]. At that time it became clear that the MCE is a phenomenon occurring in all magnetic materials when subjected to magnetic field variations, originating from the energy/entropy exchange between the magnetic and atomic lattice reservoirs. Through the fundamental Maxwell relations, the MCE can be expressed as the adiabatic temperature (ΔT_{ad}) and the isothermal magnetic entropy (ΔS^{m}_{iso}) variation that a magnetic material undergoes when varying an external magnetic field [3]:

$$\Delta T_{ad}(S,\Delta H)_{\Delta H} = -\mu_0 \int_{H_0}^{H_1} \left(\frac{T}{C(T,H)} \right)_H \left(\frac{\partial M(T,M)}{\partial T} \right)_H dH \qquad (1)$$

$$\Delta S^{m}_{iso}(T,\Delta H)_{\Delta H} = \mu_0 \int_{H_0}^{H_1} \left(\frac{\partial M(T,M)}{\partial T} \right)_H dH \qquad (2)$$

where T is the temperature of the material, $\Delta H = H_1 - H_0$ is the magnetic field variation, C is the heat capacity and M the magnetization. Hence, in order to maximize ΔT and ΔS, one important parameter to consider is the magnetization variation as a function of temperature: this means that the MCE reaches its maximum at the magnetic transition temperatures (the Curie temperature T_C, or the Néel temperature T_N) for all magnetic materials and consequently the operational temperature range of each material is delimited by a small interval close to its magnetic transition temperature. Furthermore, the MCE maximization depends strongly on the value of the magnetization derivative with respect to temperature $\partial M/\partial T$, which is specific for each material and depends on the order of the associated phase transition.

Still in the 1930s, Giauque understood that he could make use of the large magnetization change at low temperatures of a paramagnetic salt (particularly gadolinium sulphate) to reach sub-Kelvin temperatures for the first time [4] and for a long time magnetic refrigeration was only considered for low temperature applications. Forty years later, in 1973, Brown [5] used gadolinium (which has a T_C around 295 K) to design a magnetic refrigerator capable of operating at room temperature [6,7]. However, the real drive towards room temperature magnetic refrigeration occurred with the aforementioned events during the 1980s [1]. Finally, in 1997, Pecharsky and Gschneidner discovered the reversible giant magnetocaloric effect (GMCE) in the compound $Gd_5Si_2Ge_2$ [8], renewing the optimism in this technology and attracting the attention of materials scientists from different areas to its study. They indirectly measured an entropy change of 18 J Kg^{-1} K^{-1} at 278 K for a field change of 0–5 T [8]. Later, Morellon and co-workers [9] discovered that the origin of such a giant effect is the simultaneous magnetic and structural (magnetostructural) transition that $Gd_5Si_2Ge_2$

undergoes at 278 K and that can also be triggered by the application of a magnetic field. The magnetostructural transition ensures a first-order nature of the magnetic transition and it is the reason for the giant $\partial M/\partial T$ value and consequently for the maximization of ΔS^m. Moreover, since a magnetostructural transition occurs, it is important to consider the additional entropy change contribution due to the structural transition, ΔS^{str} [10]. Since ΔS^m accounts for the purely magnetic driven processes, the total entropy variation is: $\Delta S^t = \Delta S^m + \Delta S^{str}$. Because of the reasons mentioned above it became clear that first-order magnetostructural transitions are a requirement for the GMCE, and very rapidly several materials were found that fulfill this requirement, such as NiMnGa [11]; NiMnInCo [12]; MnCoGeB [13], MnFePGe [14] and MnAs [15]. There are other materials presenting GMCE, that do not exhibit any structural transition (in the sense that their symmetry group does not change), yet still undergo large simultaneous changes in their unit cell volume and in their magnetization, as for instance $La(FeSi)_{13}$ [16], $La(FeSi)_{13}H$ [17] and $LaCaMnO_3$ [18]. It was found that materials undergoing magnetostructural/magnetovolume transitions provide other giant magneto-responsive properties with technological interest besides the giant magnetocaloric effect, such as giant magnetostriction [19] and magnetoresistance [20]. It is important to remark that for these materials the magnetostriction is not a consequence of the reorientation of magnetic domains, as in the classical sense of magnetostriction, but instead it is a result of the structural transition induced by a magnetic field, which gives rise to sharp changes in the lattice parameters and consequently to macroscopic shape and/or volume changes of the material.

Thus, magnetostructural transitions have been attracting an increasing number of scientists, from both experimental and theoretical backgrounds, due to the multifunctionality arising from these effects. In systems exhibiting magnetostructural transitions there is a strong competition between the magnetic and crystallographic degrees of freedom involved in the transition, which is very sensitive to extrinsic parameters—such as temperature, magnetic field, pressure—and intrinsic properties—disorder (chemical and strain) and microstructure [1,21–24]. It is very difficult to theoretically predict the influence of intrinsic properties, but the effect of external parameters, such as magnetic field and temperature, can be predicted. In 2006, Paudyal and co-workers [25] developed a pioneering work based on the fundamental thermodynamic equations and on results from first principles calculations, which allowed the theoretical estimation of the temperature dependencies of several magnetic properties (including the MCE) of the $Gd_5Si_2Ge_2$ compound. They were able to plot the free energies of both structures in competition as a function of temperature and from the behavior of these curves they were able to estimate whether there is a magnetostructural transition, as well as the temperature at which it occurs. Recently, the same approach was extended to the $Tb_5(Si,Ge)_4$ system allowing a better understanding on the (de)coupled nature of the magnetic and structural transitions [26,27]. Moreover, first principles calculations have also shown to be a fundamental tool to understand the correlation between electronic and magnetic interactions within the unit cell of the $R_5(Si,Ge)_4$compounds [26,28].

It is noteworthy that whether there is a magnetostructural transition or not is a critical feature both from the scientific point of view, in order to understand the origins of such large magnetic responses such as the magnetocaloric, magnetoresistive and magnetostrictive effects, but also technologically, since the occurrence of a magnetostructural transition promotes a giant enhancement on the magnitude of all these effects and specially the MCE. This statement is very patent in the Gschneidner and co-workers report were they show that ΔS^{str} can account for more than half of the total entropy variation for applied magnetic fields below 2 T in $Gd_5(Si,Ge)_4$ compounds and its contribution can be even higher for other compounds. This means that for materials where no magnetostructural transition is observed, in principle it should be possible to greatly optimize their MCE through the strengthening of their magnetic and structural coupling. Simultaneously, it also means that for materials where a magnetostructural transition is observed, the path to optimize their MCE should not be focused in increasing their magnetic and structural coupling.

In this review we will focus on the nature of a first order magnetic transition from the point of view of a competition between two crystallographic phases. The thermodynamic framework will be introduced in combination with first principles calculations. Afterwards we will discuss the importance of knowing the magnetic transition temperatures of both phases involved and a general method to estimate their values through simple experimental measurements. The practical cases of $Gd_5Si_2Ge_2$ and $Tb_5Si_2Ge_2$ will be presented to explain the (de)coupling of the magnetic and structural phase transitions in both compounds. By incorporating the aforementioned external parameters, the plots of the free energy as a function of temperature will be presented and the effect of an applied magnetic field will be discussed. Finally, the correlation between this phase competition and one exotic phenomenon—the Griffiths-like phase—will also be discussed and presented as a fingerprint of systems with strong competition between phases.

2. THERMODYNAMICS OF FIRST ORDER PHASE TRANSITIONS

2.1. General Approach

According to the Ehrenfest classification, a first order phase transition must exhibit a discontinuity in the entropy variation and in the order parameter which arises from a crossover between two free energies, resulting in two distinct states: a low temperature (LT) and a high temperature (HT) state. It is the change between these two states that leads to the discontinuity of the entropy and the maximization of the entropy variation, leading to a giant magnetocaloric effect in the case of magnetic materials. Besides temperature, it is crucial that the application/removal of the external parameters (pressure, magnetic field, *etc.*) can be made near the phase transition region to be able to modify the system. To understand this process thermodynamically, let us consider the free energy for a simple magnetic material [25]:

$$F(T,\sigma) = U - TS = U^{lattice} - HgJ\sigma - \frac{3}{2}\frac{J}{J+1}k_B T_C \sigma^2 - TS^m - TS^{lattice} \quad (3)$$

where σ is the reduced magnetization, the first term ($U^{lattice}$) is the non-magnetic internal energy at T = 0 K, the second term is the interaction energy between a localized magnetic moment J and the applied magnetic field H (Zeeman term), the third one is the magnetic ion-ion exchange interaction energy and the last two terms are the entropic contributions (magnetic and lattice, respectively) to the system's free energy.

With the help of first principles calculations, one can estimate the free energy of any structure and, in particular, of any two structures (say 1 and 2) involved in a structural transition 1→2 (where 1 and 2 correspond to the HT and LT magnetization states, respectively): ΔF = F[1] − F[2]. This will be equal to the difference between the $U^{lattice}$ of both structures at T = 0 K and H = 0 T, which are estimated from first principles calculations. Considering the mean field theory, the magnetic entropy per magnetic ion (S_M) for a general J can be estimated from:

$$S_M(\sigma) = k_B \ln\left(\frac{\sinh\left(\frac{2J+1}{2J}B_J^{-1}(\sigma)\right)}{\sinh\left(\frac{1}{2J}B_J^{-1}(\sigma)\right)}\right) - \sigma B_J^{-1}(\sigma) \sim k_B\left[\ln(2J+1) - \frac{3}{2}\frac{J}{J+1}\sigma^2 + O(\sigma^4)\right] \quad (4)$$

where B−1JBJ−1is the inverse Brillouin function. The magnetic equation of state ($\sigma(T,H)$) can be obtained by finding the zero of the derivative of Equation (3) with respect to σ [leading to $\sigma = \sigma(T,H)$], which corresponds to the condition of minimum free energy. This minimum free energy is then given by $F^{min}(T,H) = F(T, \sigma(T,H))$. From Equation (3) one can see that the free energy difference between structures 1 and 2 arises from a) the non-magnetic contribution ($F_{lattice} = U^{lattice} - TS^{lattice}$) and b) from the exchange and magnetic entropy terms (2nd, 3rd and 4th terms in Equation (3)), given the hypothetically different T_C of the two structures. Therefore it becomes critically important to know the T_C of each structure involved in order to be able to plot the free energies of both structures as a function of temperature. Such a plot will allow one to understand whether or not there is a structural transition occurring and, if so, at what temperature it should occur. The first obstacle in this approximation is the determination of the magnetic ordering temperatures (T_C^1 and T_C^2) of both structures involved in a phase transition.

2.2. How to Inspect the "Hidden" Magnetic Ordering Temperatures

In standard materials exhibiting second-order magnetic transitions, their magnetic long range ordering sets in at a temperature that can be easily assessed by measuring their magnetization as a function of temperature at constant magnetic field, analysing its derivative and signalling the temperature at which a maximum/minimum occurs. Unfortunately, such a simple approach is not appropriate in systems presenting magnetostructural transitions, since in these systems the structural transition occurs somewhere in between T_C^1 and T_C^2, *i.e.*, $T_C^1 < T_S < T_C^2$. Let us consider the M(T) measurement on cooling of such an ideal system. Starting with a low magnetization value associated with the paramagnetic state of structure 1, there will be no significant changes on M(T) when cooling through T_C^2. Then, when $T = T_S$, a sharp increase towards a much higher magnetization value will be observed, which is already associated with the ferromagnetic behavior of structure 2. While continuing on cooling down the system, no major change will be observed in the M(T) curve, not even when crossing the lower Curie temperature of structure 1, T_C^1, because the system is already transformed into the structure 2. In conclusion, the T_C^1 and T_C^2 will be invisible to the M(T) measurement and that is why we have to figure out an alternative method to estimate T_C^1 and T_C^2.

In order to be direct and clear, two practical examples will be considered. Both are materials presenting giant magnetocaloric effect: $Gd_5Si_2Ge_2$—with a simultaneous magnetic and structural transition—and $Tb_5Si_2Ge_2$—presenting non-simultaneous magnetic and structural transitions. Particularly for these two materials, the structures 1 and 2 mentioned above correspond to the monoclinic (M) and orthorhombic [O(I)] structures, whereas the two magnetic states considered are paramagnetic (PM) and ferromagnetic (FM). The remaining relevant information is summarized in Table 1.

Table 1. Relevant information about $Tb_5Si_2Ge_2$ and $Gd_5Si_2Ge_2$ compounds.

	$Tb_5Si_2Ge_2$	$Gd_5Si_2Ge_2$
Simultaneous transition	×	√
Structure 1/2	M/O(I)	M/O(I)
Curie temperatures, T_C (K)	112/200 [27,29]	208/305 [25,27]
Structural transition temperature, T_S(K)	95 [29]	270 [30,31]

Although these compounds were already thoroughly studied [8,23,32,33], in order to exemplify the broad range of application of this procedure, we will only take into account the fact that both materials undergo structural transitions under temperature and/or magnetic field changes. We will also assume that for both structures involved, the FM state is the most stable at low temperatures. Hence, with these assumptions, there are four different possible states: $[M;PM^M]$, $[M;FM^M]$, $[O(I);PM^{O(I)}]$ and $[O(I);FM^{O(I)}]$ at any temperature and field value. Now, let us consider that the temperature of the material ($Tb_5Si_2Ge_2$ or $Gd_5Si_2Ge_2$) is slightly above its structural transition temperature T_S, where the material stabilizes at its $[M;PM^M]$ state. At this temperature it is possible to induce the magnetostructural transition, *i.e.*, the transition from the $[M;PM^M]$ state to the $[O(I);FM^{O(I)}]$

state through the application of a strong enough magnetic field (critical field, H_C). Hence, for high magnetic fields ($H > H_C$) it is possible to inspect the magnetic behavior of the O(I) structure, even at $T > T_S$. In fact, as expected, the critical field H_C depends on the temperature of the material and conversely the structural transition temperature T_S depends on the applied magnetic field intensity: $T_S(H) \neq T_S \equiv T_S(H = 0)$. Experimentally, such behavior can be more clearly understood in the isothermal magnetization curves as a function of applied field for several different temperatures (below and above T_S). Two clearly distinct types of curves will be obtained: for $T < T_S$ isotherms, a typical FM steep increase in the magnetization up to the saturation field is observed, whereas for $T > T_S$ the magnetization exhibits two regimes: a low magnetization (LM) state at low fields undergoes a sharp metamagnetic transition into a high magnetization (HM) state at $H > H_C$ as a consequence of the very different magnetic susceptibilities of structures M and O(I). The data are then represented in the form of Arrott plots [34]—H/M as a function of M^2-, as can be seen in Figures 1a,b for $Gd_5Si_2Ge_2$ and $Tb_5Si_2Ge_2$, respectively. This representation is based on the Landau theory of phase transitions and follows from the magnetic state equation [34]:

$$\frac{H}{M} = \alpha(T - T_C) + BM^2 \quad (5)$$

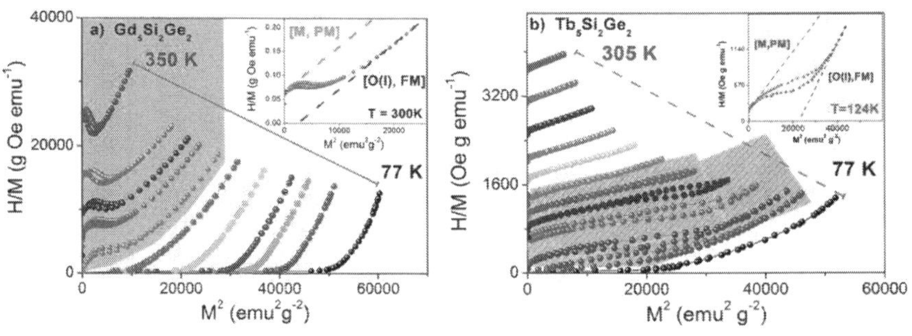

Figure 1. Arrott plots for $Gd_5Si_2Ge_2$ (**a**) in the temperature range 77–350 K; Inset: T = 300 K, and for $Tb_5Si_2Ge_2$ (**b**) in the temperature range of 77–305 K; Inset: T = 124 K.

From this expression one recognizes that when plotting the data in the H/M *vs.* M^2 representation, linear curves are expected, intercepting the positive y-axis when $T > T_C$ (in the PM state) and the positive x-axis when $T < T_C$ (in the FM state). In this representation, the critical isotherm ($T = T_C$) is a straight line that goes through the origin. The intercept with the positive y-axis, $\alpha(T-T_C)$, is the inverse susceptibility $\chi^{-1}(T)$. Let us consider the $Gd_5Si_2Ge_2$ case. As can be seen in Figure 1a, the curves in the green region exhibit two different linear regimes. This is a signature that they belong to the $T_S(H) > T > T_S(H = 0)$ temperature range, where the magnetic field applied is able to induce the structural transition $[M,PM^M] \rightarrow [O(I),FM^{O(I)}]$. M has lower magnetic susceptibility (smoother slope) than the O(I) structure (steeper slope) (inset of Figure 1a). Now, the squared spontaneous magnetization $M_S^2[O(I)]$ of the O(I) structure (or HM state) can be estimated as the interception of the linear extrapolation of the high field magnetization data with the abscissa axis at a fixed temperature. In the white region—$T < T_S$ temperature range—only one linear regime is observed because the system is already in the $[O(I), FM^{O(I)}]$ state. By following the same procedure, the $M_S^2[O(I)]$ values at lower temperatures are estimated. By combining the $M_S^2[O(I)]$ values obtained for $T < T_S$ and for $T > T_S$, it is possible to plot M2sMs2 [O(I)] as a function of temperature as depicted in Figure 2a (blue dots). Simultaneously, the inverse paramagnetic susceptibility of the M structure (LM state) can also be determined from the low field isothermal curves ($T > T_S$) and their intercept with the positive y-axis, yielding $\chi^{-1}[M](T)$, which is also displayed in Figure 2a) (red dots). Finally, by fitting the M2sMs2 [O(I)](T) and the $\chi^{-1}[M](T)$ curves with Brillouin and linear functions, respectively, it is possible to estimate the Curie temperatures of both M and O(I) phases. In this example the fits result in the following critical temperatures for O(I) and M structures: $T_C^{O(I)} \sim 308$ K and TMC~251TCM~251 K, in accordance with previous results [25,27,35,36]

The Arrott plots for $Tb_5Si_2Ge_2$ are shown in Figure 1b. As in the $Gd_5Si_2Ge_2$ plots, the green area highlights the temperature range where the magnetic field is able to induce the M to O(I) phase transition. At T = 124 K (inset of Figure 1b), the behavior of the curves signals the magnetostructural transition from the LM $[M, PM^M]$ to the HM $[O(I), FM^{O(I)}]$ state. Following the same procedure as in the $Gd_5Si_2Ge_2$ case, the data analysis allows extracting the temperature dependence of both the inverse paramagnetic susceptibility of the M structure (red dots in Figure 2b) and the spontaneous magnetization of the O(I) structure (blue dots in Figure 2b). Again, by fitting these data, the extracted values are: TO(I)C~200TCO(I)~200 K and TMC~112TCM~112K, where the latter one is coincident with the T_C ~ 112 K obtained from the maximum derivative of the experimentally measured M(T) curves [29].

In summary, the general procedure to extract the "hidden" magnetic ordering temperatures of structures 1 and 2 is as follows:

1) Measure the magnetization as a function of magnetic field for several temperatures in the range where T_S is expected (a broader interval will provide more points and hence more accurate fits), ensuring that for some isotherms the field is strong enough to induce the metamagnetic transition;

2) Replot the data in the Arrott form (H/M as a function of M^2) and identify the two different linear regimes associated with LM (1) and HM (2) states for $T > T_S$;

3) Perform linear fits to these two different regimes and estimate their interceptions with the positive y-axis (for the LM state) and with the positive x-axis (for the HM state) in order to extract, respectively, the inverse susceptibility of the LM (1) state—χ^{-1}(LM)—and the squared spontaneous magnetization of the HM (2) phase—M^2_S (HM).;

4) Plot χ−1(1) and M2sMs2 (2) as a function of temperature;

5) Through linear and Brillouin fits to these curves, the magnetic ordering temperatures of both 1 and 2 structures are extracted.

To the best of our knowledge, this procedure has only been applied to the $R_5(Si,Ge)_4$ family of compounds. However, it could be useful for studying several other materials displaying magnetic field-induced structural transitions, hence with extremely sensitive magneto-responsive properties that are attractive for technological applications. Examples of such materials are the multifunctional magnetic shape memory alloy $Ni_{55}Mn_{20}Ga_{25}$ [11], the MnNiFeGe family [12], MnCoGeB [13], MnFePGe [14], and the materials undergoing giant volume changes within the same crystallographic structure: La-Fe-Si [16] and $LaCaMnO_3$ [18].

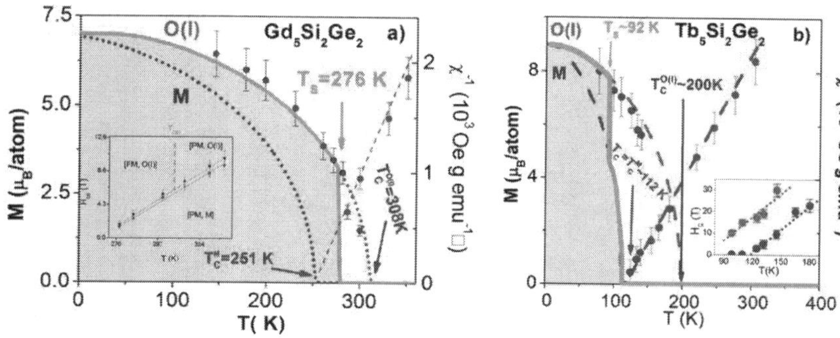

Figure 2. Spontaneous magnetization (blue dots) and reciprocal susceptibility (red dots) curves of: (**a**) $Gd_5Si_2Ge_2$ and (**b**) $Tb_5Si_2Ge_2$, obtained from the Arrott plots. The dark green dotted lines are the Brillouin fits to the spontaneous magnetization of the O(I) structure for each compound which enable the estimation of their T_C: $T_C^{O(I)} = 308$ K for $Gd_5Si_2Ge_2$ and $T_C^{O(I)} = 200$ K for $Tb_5Si_2Ge_2$. The red dashed lines are the linear fits to the reciprocal susceptibility data, which allowed estimating the T_C^M for each compound: $T_C^M = 251$ K for $Gd_5Si_2Ge_2$ and $T_C^M = 112$ K for $Tb_5Si_2Ge_2$, whereas the purple dotted lines represent the Brillouin curves assuming the obtained TMCTCM. The gray zones represent the FM regions and their boundaries (represented by the light-green curves) schematically represent the experimentally measured M(T) curves. Insets: temperature dependence of the critical magnetic fields ($H_C(T)$) obtained from the maximum of dM/dH.

2.3. Free Energy Crossings of Real Systems

Once the T_C of both structures is estimated from the Arrott plots using the method described above, their free energies can be calculated as a function of temperature. Then, the temperature at which the free energy curves of both structures intersect will define the structural transition temperature T_S. As a first approximation, we neglect the lattice entropy term and consider the magnetic contribution to the free energy in Equation (3), as well as the result of first principles calculations that yields $F_{min}^{lattice}(T=0K) = U^{lattice}$ (the first term in Equation (3)). In order to normalize the energy range to zero, the value of the free energy at 0 K of structure 2 is subtracted from the free energy FlatticeminFminlattice of both structures, resulting in: $\Delta F_{min}(T)[1,2] = F_{min}(T)[1,2] - F_{min}^{lattice}(T=0K)[2]$ Let us start with a zero magnetic field, meaning that the second term in

Equation 3 is also zero. The temperature dependence of the magnetic free energies $\Delta F_{min}[1]$ and $\Delta F_{min}[2]$ (that start at zero because of the subtraction mentioned above) are represented for the two examples $Gd_5Si_2Ge_2$ and $Tb_5Si_2Ge_2$ inFigures 3a,b, respectively. We recall that structure 1 corresponds to the monoclinic (M) crystallographic phase and structure 2 corresponds to orthorhombic O(I) phase. The first principles calculations were performed using Wien2K code, via the augmented plane wave method with local orbitals (APW + lo) and using the local spin-density approximation method allowing a correct estimation of the free energies of the M and O(I) phase at T = 0 K.

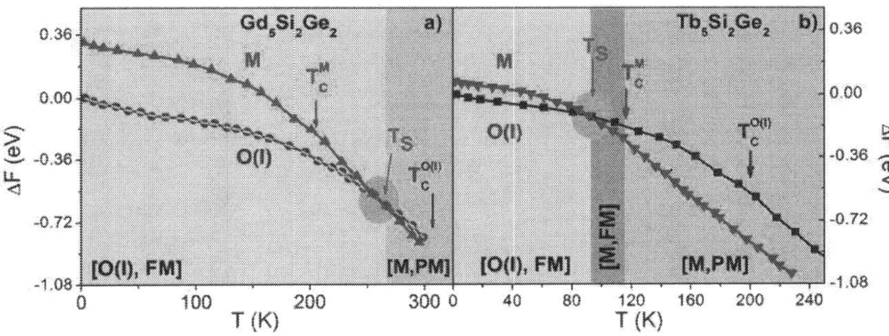

Figure 3. Free energy as a function of temperature considering the magnetic entropy and the first principles calculations ($U^{lattice}$) for both M and O(I) structures in the (**a**) $Gd_5Si_2Ge_2$ and (**b**) $Tb_5Si_2Ge_2$ materials. T_S is defined as the temperature at which the crossing $\Delta F_{min}[M] = \Delta F_{min}[O(I)]$ occurs.

It can be clearly seen that in both cases there is an intersection of the $\Delta F_{min}[O(I)]$ and $\Delta F_{min}[M]$ curves at a temperature $T = T_S$, *i.e.*, the structural transition temperature for each system. In the $Gd_5Si_2Ge_2$ case (Figure 3a), $T_S \sim 265$ K lies right between the two magnetic transition temperatures, $T_C^M < T_S < T_C^{O(I)}$, which means that upon cooling the system undergoes simultaneously a magnetic—paramagnetic to ferromagnetic $[PM^M] \rightarrow [FM^{O(I)}]$—and a structural transition from a monoclinic to an orthorhombic O(I) structure $[M] \rightarrow [O(I)]$. It is worth noting that, magnetically, the system goes from the monoclinic paramagnetic state to the orthorhombic ferromagnetic state, *i.e.*, there is a large discontinuity on both the atomic and lattice parameters and the magnetization values and these two transitions are coupled, meaning that a magnetostructural transition [M, PM^M]→[O(I), $FM^{O(I)}$] takes place. In the case of $Tb_5Si_2Ge_2$, $T_S < T_C^M < T_C^{O(I)}$, meaning that on cooling the system first undergoes the [PM^M]→[FM^M] magnetic transition at $T_C^M \sim$ 112 K, closely followed by the M→O(I) structural transition a few kelvins below, at $T_S \sim$ 95 K, *i.e.*, the two transitions are decoupled. In contrast with $Gd_5Si_2Ge_2$, here the magnetic transition is from the PM to the FM state of the monoclinic structure. The T_S values estimated with this theoretical model corroborate the values obtained from experimental measurements (particularly X-ray diffraction as a function of temperature) for these two compounds [29,30]. This demonstrates that such theoretical procedure can predict structural transition temperatures and their coupled/decoupled nature. As mentioned in the introduction section, this information is of key importance, since it allows evaluating the potential of the material regarding the optimization of its MCE. In particular, from the results of the two cases presented here it becomes clear that: (1) the MCE in $Tb_5Si_2Ge_2$ can still be greatly increased, whereas (2) $Gd_5Si_2Ge_2$ MCE cannot be improved much further by the further enhancement of its magnetic and structural coupling. Two of the most common ways to enhance this coupling is by applying pressure and by alloying. So far, our predictions have been confirmed by several works: concerning: (1), an MCE enhancement of more than 40% was reported by promoting the fully coupled magnetic and structural transitions, both by alloying with Fe [37] and by applying pressure [24], whereas regarding (2), the MCE was not improved (in fact it was smaller) when the same alloying and pressure were applied [38,39].

The $T_C^M < T_S < T_C^{O(I)}$ scenario, or more generally $T_C^1 < T_S < T_C^2$, is more attractive from the magnetic entropy change maximization point of view. As previously discussed by Liu [40], Pecharsky, Gschneidner and co-workers [21,41], the magnetic entropy change peak (ΔSmmaxΔSmaxm) will be higher for systems where simultaneously $\Delta T_C = T_C^2 - T_C^1$ is maximized and T_S is just slightly above T1CTC1. In this scenario, the magnetization will jump (ΔM) from a nearly zero value to almost the maximum value of the magnetization of structure 2 (ΔM will be proportional to the difference $\Delta T_{CS} = T_C^2 - T_S$), hence maximizing the ∂M/∂T at T_S and consequently the ΔSmmax.

Now, let us turn on the magnetic field. As can be seen in Equation (3), the second term (*HgJσ*) is always positive, but it is preceded by a minus sign. Therefore, the effect of a non-zero magnetic field is to lower the free energy. However, due to the different T_C on each crystal structure (that critically influence the magnetization $\sigma(T_C)$), the free energy of the structure with higher T_C (O(I), in both materials) decreases more rapidly than the one with lower T_C (M). In this way the application of a magnetic field tends to increase the temperature at which the free energy crossover occurs—*i.e.*, where $\Delta F_{min}[O(I)] = \Delta F_{min}[M]$—as can be confirmed in Figure 4. Therefore, the structural transition temperature $T_S(H)$ is increased. This was confirmed experimentally and a linear dependence of the critical magnetic field as a function of temperature $H_C(T)$ was observed, as shown in the insets of Figure 2a,b.

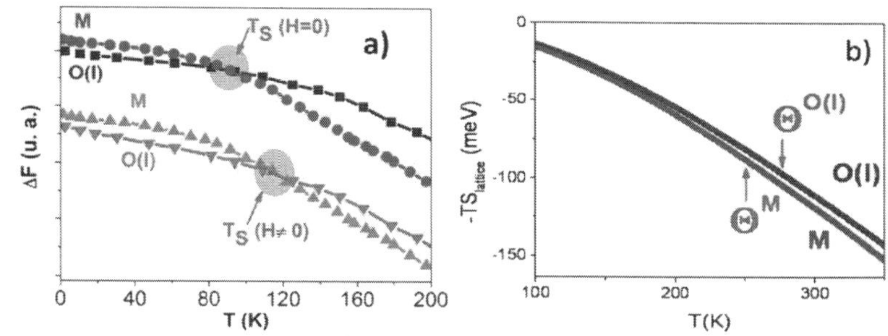

Figure 4. (**a**) Schematic free energy of M and O(I) structures as a function of temperature for the zero (upper pair of curves) and non-zero (lower pair of curves) magnetic field; (**b**) Lattice contribution to the free energy of M and O(I) structures of $Gd_5Si_2Ge_2$ compound as a function of temperature.

So far, the effect of $F^{lattice} = TS^{lattice}$ was still not considered in this analysis. To analyze the contribution of the entropic lattice term, the Debye approximation is used. The entropy per ion is given by the following expression [40]:

$$S^{lattice}(\Theta) = 3k_B \ln\left(1 - e^{\frac{-\Theta}{T}}\right) + 12k_B\left(\frac{T}{\Theta}\right)^3 \int_0^{\frac{\Theta}{T}} \frac{x^3}{e^x - 1}dx$$
$$\sim 3k_B \ln\left(\frac{\Theta}{T}\right) + \frac{3}{40}k_B\left(\frac{\Theta}{T}\right)^2 + O\left[\left(\frac{\Theta}{T}\right)^3\right] \qquad (6)$$

where Θ is the (cut-off) Debye temperature. From this expression one can infer that the high temperature phase entropic contribution to the lattice free energy ($-TS^{lattice}$) is lower for the structure with the lower Θ. In fact, the Debye temperature for the M phase is lower than of the O(I) phase. This can be confirmed from the relation $\frac{\Theta^{M}}{\Theta^{O(I)}}=\left(1-\eta\frac{\Delta V}{V}\right)$, where η is the Grüneisen parameter (typically around 10 for $Gd_5Si_2Ge_2$) and $\frac{\Delta V}{V}$ is the relative volume increase when the system goes through a O(I) to M structural change (typically ~1%). An estimation of Θ^M considering the experimentally detected velocity of sound in $Gd_5Si_2Ge_2$ resulted in $\Theta^M \sim$ 250 K [42] and consequently $\Theta^{O(I)}=\frac{\Theta^{M}}{\left(1-\eta\frac{\Delta V}{V}\right)}=278$ K. The lower Θ for the M phase was also experimentally confirmed in the case of $Y_5(Si_xGe_{1-x})_4$ system [43]. Furthermore, with the estimated Θ for each structure of $Gd_5Si_2Ge_2$ (Θ^M and $\Theta^{O(I)}$) it is possible to plot the lattice contribution from each structure to the free energy. This is displayed in Figure 4b. It can be seen in this figure that at high temperature the entropic lattice contribution also tends to stabilize the M phase, as shown for the $Y_5(Si_xGe_{1-x})_4$ system in [44]. Therefore it is clear, at least for these two examples, that including the lattice contribution ($F^{lattice} = TS^{lattice}$) will not drastically affect the temperature dependence of the total free energy of each structure and, consequently, the main findings about the (de)coupled transition temperature discussed above.

However, the entropy change associated with the structural transition (ΔS^{str}) cannot be ignored, as it plays a major role on the total entropy variation (ΔS^{T}) of the majority of the systems undergoing magnetic field-induced structural transitions. As Pecharsky and Gschneidner [41] noticed, the lattice contribution to the total entropy change, ΔS^{str}, varies from system to system because it depends on the entropy differences between both the initial and final structures. They also show that $\Delta S^{str}$is proportional to the volume difference between the two structures. Two examples are given: ΔS^{str} is around 50% of ΔS^{T} for the $Gd_5Si_{2.09}Ge_{1.91}$ compound and about 93% of ΔS^{T} for $MnCoGeB_{0.2}$ under a magnetic field change of 5 T [41]. Furthermore, whereas the magnetic entropy change (ΔS^{m}_{max}) increases with $H^{2/3}$ [45], the relation between ΔS^{str} and H is harder to unveil. However, for these two systems a magnetic field change of 5 T is enough to complete the structural transition and hence ΔS^{str} reaches its saturation value.

3. IS THE GRIFFITHS-LIKE PHASE A REQUISITE FOR A GIANT MAGNETOCALORIC EFFECT?

Together with the strong magneto-responsive properties, magnetic materials presenting strong coupling between structural and magnetic properties exhibit other exotic effects, such as the appearance of Griffiths-like phases. After the model proposed by Griffiths in the 1960s, the presence of Griffiths-like phases in several real materials has been recently reported [46]. In his model, Griffiths came upon with a diluted Ising ferromagnetic system with a well-defined magnetic ordering temperature T_C which was obviously lower than the magnetic ordering temperature of the undiluted (or pure) ferromagnetic system (T_G), *i.e.*, $T_C < T_G$. He was able to show that for $T_C < T < T_G$ it was possible that the reciprocal magnetic susceptibility of the system could deviate from its expected Curie-Weiss linear behavior. In fact, this deviation would start, on cooling, exactly at T_G, as if the diluted system somehow "remembered" that its magnetic ordering temperature was T_G "before" it became diluted [33]. Therefore, the concept of a Griffiths phase, also named Griffiths singularity, is based on the emergence of a new magnetic system within the paramagnetic matrix state, which evidences ferromagnetic short-range ordering [46–48]. Although these short-range clusters would not reveal themselves in the typical M(T) curves, they would be responsible for slight deviations from the linear behavior of the reciprocal magnetic susceptibility as a function of temperature.

In the past few years several R_5T_4 compounds (R = rare earth: Gd, Tb, Dy, and Ho; T = group 14 elements) have shown a Griffiths-like behavior [33,48–50]. Such phenomenon was firstly reported in the $Tb_5Si_2Ge_2$ compound, when it was associated with the experimentally observed deviation from the expected linearity (from Curie-Weiss law) of the χ^{-1} (T) curve in the paramagnetic region, $T > T_C$ [33].

Later, different works, reported by Ouyang, Tian, Pereira and co-workers identified the Griffiths-like phase in other compounds of the R_5T_4 family [23,48–50]. All these compounds presented anomalies featuring their reciprocal susceptibility curve: a "stair-like" behavior occurring in an intermediate temperature region, $T_C < T < T_G$. Later on, our group has completed the phase diagrams of the $R_5(Si_xGe_{1-x})$ (R = Gd, Tb, Dy and Ho) compounds with their corresponding Griffiths-like phase [50]. It was observed that despite the different main temperatures along their composition for each R compound (T_G, T_C and T_N), the phase diagrams were very similar in shape and two main conclusions could immediately be drawn: (1) the Griffiths singularity behavior appears only in the M and O(II) phases for compositions below a characteristic Si concentration that was defined as x_p and (2) the O(II) structure is stabilized when the ratio $\tau \equiv T_{C,N}/T_G$ is ~ 0.5, regardless of the rare-earth element. The third and more general conclusion was reached when these diagrams were replotted as τ as a function of $(x{-}x_p)$ and the collapse of all four phase diagrams into a single curve was observed, as represented in Figure 5. The physical mechanism behind the appearance of the Griffiths-like phase is the same, regardless of the rare-earth element, and it is deeply associated with the Si/Ge ratio. In fact, the Ge substitution of Si makes a perfect analogy for the original undiluted/diluted system Griffiths idealized, *i.e.*, when no Ge is present or its concentration is such that $x > x_p$ (x_p will be defined below), the system stabilizes in the O(I) structure where all interslab bonds are formed. A good review on the particular structural features of this system is given by Miller [1]. In the O(I) structure there is a strong and collinear ferromagnetic state with the highest T_C for each phase diagram—the system can be thought of as undiluted/pure. When the Ge substitution leads to a Si critical concentration (x_p is defined as this Si concentration value) the interslab bonds start to break (it is hypothesized that it is when $x = x_p$ that the Ge atoms start substituting Si atoms at the interslab sites) and the O(II) or M structures become the most stable structures, with a simultaneous decrease in T_C. In these conditions the system can be thought of as diluted. It is well-known from previous studies that the magnetism in these compounds exists from the competition of ferromagnetic (within each slab) and antiferromagnetic interactions (mostly between slabs). More accurately, at the interslab region, the main magnetic exchange mechanism is the Ruderman-Kittel-Kasuya-Yosida (RKKY) exchange. The RKKY magnetic exchange energy parameter J_{RKKY} is a Bessel function which starts at a positive value (ferromagnetic interaction) but decays fast with increasing distance between the atomic moments, crossing zero and becoming negative (antiferromagnetic interaction) at a critical distance. Therefore, when Ge starts replacing Si at the interslab sites (at x_p), and since it has a higher atomic radius, it promotes an increase in the distance between consecutive slabs. Because of the RKKY interaction, this leads to a decrease of the magnetic exchange energy and consequently to decrease in T_C. When a critical Ge concentration limit, x_C, is reached, the RKKY interaction becomes negative and hence the AFM state becomes the most stable one. The x_C is the ferromagnetic percolation limit for these diluted systems, below which it is not possible to achieve long range ferromagnetic order. Finally, the complete analogy with the Griffiths diluted ferromagnetic system is schematized inFigure 5b.

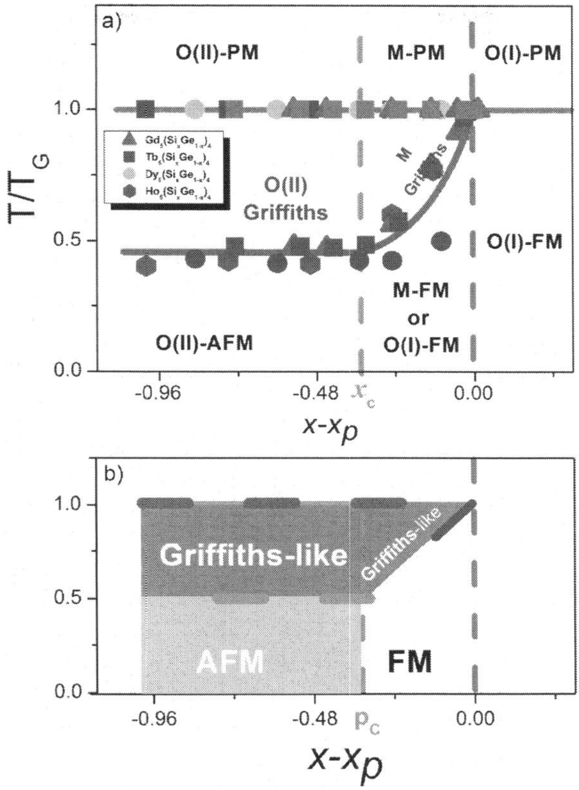

Figure 5. (**a**) Universal x,T phase diagram of the $R_5(Si_xGe_{1-x})_4$ compounds with R = Gd (triangles), Tb (squares), Dy (circles), and Ho (hexagons); (**b**) Adapted T *vs.* (x − x_p) phase diagram of a dilute FM system for the $R_5(Si_xGe_{1-x})_4$ system [50].

With this set of comprehensive works the extent of the Griffiths-like behavior on the $R_5(Si,Ge)_4$ compounds is shown, highlighting their magnetic nature as that of a diluted ferromagnetic system where the Ge substitution is the dilution mechanism—and hence demonstrating why the $R_5(Si,Ge)_4$ are an ideal case study for the Griffiths phase phenomenon. In addition it was found that in all the $R_5(Si,Ge)_4$ materials presenting a Griffiths-like phase [$Gd_5(Si,Ge)_4$; $Tb_5(Si,Ge)_4$; $Dy_5(Si,Ge)_4$ and $Ho_5(Si,Ge)_4$], T_G is systematically equal to the T_C of the O(I) phase. This has led to the conclusion that in the $R_5(Si,Ge)_4$ compounds, the low temperature phase can easily be assessed through simple measurements of magnetization as a function of temperature at constant magnetic field. In summary, it can be stated that that the $R_5(Si,Ge)_4$ compounds exhibiting Griffiths-like phase are most likely to present strong competition between two phases and consequently the Griffiths-like phase constitutes a fingerprint for strong magneto-responsive properties in this family of compounds. In fact, a rapid review of the literature allows us to identify several other strongly-responsive magnetic materials, such as transition metal oxides (particularly manganites [51,52], cuprates [52] and ACr_2X_4 spinels [53]) which exhibit colossal magnetoresistance and magnetoelasticity [54], magnetic semiconductors [55] and giant magnetocaloric compounds [33], all of which exhibit the Griffiths-like phase. This outcome allows us to suggest that the Griffiths-like phase can be a consequence of the strong correlation between magnetic and atomic lattices and of the strong competition between states in other materials beside the $R_5(Si,Ge)_4$ compounds. However, further investigation should be done in order to confirm this suggestion.

4. CONCLUSIONS

In summary, in this work we have demonstrated the importance of the strong competition between phases on the origin of several giant magneto-responsive effects, particularly the magnetocaloric effect. Such competition is responsible for an additional contribution to the entropy change that can be studied thoroughly through a combination of detailed measurement/analysis of the materials magnetic properties and first principles calculations. A method to analyze the Arrott plots in order to extract the hidden magnetic ordering temperatures is reviewed and generalized. Such parameters are shown to be crucial as inputs for the simple model of the free energy as a function of temperature and for a deeper understanding on the nature of the (de)coupled magnetic and structural phase transitions. Practical examples are given, namely by explaining the coupling of magnetic and structural transitions in $Gd_5Si_2Ge_2$ or their decoupling in the $Tb_5Si_2Ge_2$case. Finally, the Griffiths-like phases are reviewed in the $R_5(Si,Ge)_4$ families and a correlation between their universal phase diagram and the large magneto-responsive properties is discussed.

ACKNOWLEDGMENTS

The authors acknowledge to FCT for financial support in the project PTDC/CTM-NAN/115125/2009, EXPL/EMS-ENE/2315/2013 and Projeto Norte-070124-FEDER-000070. We would also like to thank Erik Kampert, José Manue Moreira, Uli Zeitler, João Nuno Gonçalves, João Amaral, Vítor Amaral and João Bessa Sousa for the important previous discussions related with this work.

AUTHOR CONTRIBUTIONS

First two authors contributed equally to this work, particularly on the writing and on the bibliographic review. André Miguel Pereira and João Pedro Araújo were responsible for this work and supervised the manuscript. Isabel T. Gomes' help was fundamental for a more clear and improved manuscript. Armandina Maria Lima Lopes, Luis Morellón, Cesar Magen, Pedro Antonio Algarabel, Manuel Ricardo Ibarra provided important inputs on the scientific analysis of the work. All authors have read and approved the final manuscript.

REFERENCES

1. Miller, G.J. Complex rare-earth tetrelides, $RE_5(Si_xGe_{(1-x)})_4$: New materials for magnetic refrigeration and a superb playground for solid state chemistry. *Chem. Soc. Rev* **2006**, *35*, 799–813.

2. Smith, A. Who discovered the magnetocaloric effect? *Eur. Phys. J. H* **2013**, *38*, 507–517.

3. Gschneidner, K.A., Jr.; Pecharsky, V.K.; Tsokol, A.O. Recent developments in magnetocaloric materials. *Rep. Prog. Phys***2005**, *68*, 1479–1539.

4. Giauque, W.; MacDougall, D. Attainment of Temperatures Below 1° Absolute by Demagnetization of $Gd_2(SO_4)_3{\cdot}8H_2O$.*Phys. Rev* **1933**, *43*, 768–768.

5. Brown, G.V. Magnetic heat pumping near room temperature. *J. Appl. Phys* **1976**, *47*, 3673.

6. Gschneidner, K.A.; Pecharsky, V.K. Thirty years of near room temperature magnetic cooling: Where we are today and future prospects. *Int. J. Refrig* **2008**, *31*, 945–961.

7. Brown, D.R.; Dirks, J.A.; Fernandez, N.; Stout, T.B. The Prospects of Alternatives to Vapor Compression Technology for Space Cooling and Food Refrigeration Applications. *Energ. Eng* **2012**, *109*, 7–20.

8. Pecharsky, V.K.; Gschneidner, K.A. Giant Magnetocaloric Effect in $Gd_5(Si_2Ge_2)$. *Phys. Rev. Lett* **1997**, *78*, 3–6.

9. Morellon, L.; Algarabel, P.A.; Ibarra, M.R.; Blasco, J.; Garcia-Landa, B.; Arnold, Z.; Albertini, F. Magnetic-field-induced structural phase transition in $Gd_5(Si_{1.8}Ge_{2.2})$. *Phys. Rev. B* **1998**, *58*, 721–724.

10. Pecharsky, V.K.; Holm, A.P.; Gschneidner, K.A.; Rink, R. Massive Magnetic-Field-Induced Structural Transformation in Gd_5Ge_4 and the Nature of the Giant Magnetocaloric Effect. *Phys. Rev. Lett* **2003**, *91*, 197204.

11. Pasquale, M.; Sasso, C.; Lewis, L.; Giudici, L.; Lograsso, T.; Schlagel, D. Magnetostructural transition and magnetocaloric effect in $Ni_{55}Mn_{20}Ga_{25}$ single crystals. *Phys. Rev. B* **2005**, *72*, 094435.

12. Liu, J.; Gottschall, T.; Skokov, K.P.; Moore, J.D.; Gutfleisch, O. Giant magnetocaloric effect driven by structural transitions. *Nat. Mater* **2012**, *11*, 620–626.

13. Trung, N.T.; Zhang, L.; Caron, L.; Buschow, K.H.J.; Brück, E. Giant magnetocaloric effects by tailoring the phase transitions. *Appl. Phys. Lett* **2010**, *96*, 172504.

14. Trung, N.T.; Ou, Z.Q.; Gortenmulder, T.J.; Tegus, O.; Buschow, K.H.J.; Brück, E. Tunable thermal hysteresis in MnFe(P,Ge) compounds. *Appl. Phys. Lett* **2009**, *94*, 102513.

15. Wada, H.; Tanabe, Y. Giant magnetocaloric effect of $MnAs_{1-x}Sb_x$. *Appl. Phys. Lett* **2001**, *79*, 3302.

16. Shen, B.G.; Sun, J.R.; Hu, F.X.; Zhang, H.W.; Cheng, Z.H. Recent Progress in Exploring Magnetocaloric Materials. *Adv. Mater* **2009**, *21*, 4545–4564.

17. Lyubina, J.; Nenkov, K.; Schultz, L.; Gutfleisch, O. Multiple Metamagnetic Transitions in the Magnetic Refrigerant $La(Fe,Si)_{13}H_x$. *Phys. Rev. Lett* **2008**, *101*, 177203.

18. Ibarra, M.R.; Algarabel, P.A.; Marquina, C.; Blasco, J.; García, J. Large Magnetovolume Effect in Yttrium Doped La-Ca-Mn-O Perovskite. *Phys. Rev. Lett* **1995**, *75*, 3541–3544.

19. Hadimani, R.L.; Bartlett, P.A.; Melikhov, Y.; Snyder, J.E.; Jiles, D.C. Field and temperature induced colossal strain in $Gd_5(Si_xGe_{1-x})_4$. *J. Magn. Magn. Mater* **2011**, *323*, 532–534.

20. Morellon, L.; Stankiewicz, J.; García-Landa, B.; Algarabel, P.A.; Ibarra, M.R. Giant magnetoresistance near the magnetostructural transition in $Gd_5Si_{1.8}Ge_{2.2}$. *Appl. Phys. Lett* **1998**, *73*, 3462–3464.

21. Pecharsky, V.K.; Gschneidner, K.A.; Mudryk, Y.; Paudyal, D. Making the most of the magnetic and lattice entropy changes. *J. Magn. Magn. Mater* **2009**, *321*, 3541–3547.

22. Pecharsky, V.K.; Gschneidner, K.A., Jr. $Gd_5(Si_xGe_{1-x})_4$: An Extremum Material. *Adv. Mater* **2001**, *13*, 683–686.

23. Belo, J.H.; Pereira, A.M.; Araújo, J.P.; de la Cruz, C.; dos Santos, A.M.; Gonçalves, J.N.; Amaral, V.S.; Morellon, L.; Ibarra, M.R.; Algarabel, P.A.; *et al.* Tailoring the magnetism of $Tb_5Si_2Ge_2$ compounds by La substitution. *Phys. Rev. B***2012**, *86*, 014403.

24. Morellon, L.; Arnold, Z.; Magen, C.; Ritter, C.; Prokhnenko, O.; Skorokhod, Y.; Algarabel, P.A.; Ibarra, M.R.; Kamarad, J. Pressure Enhancement of the Giant Magnetocaloric Effect in $Tb_5Si_2Ge_2$. *Phys. Rev. Lett* **2004**, *93*, 137201.

25. Paudyal, D.; Pecharsky, V.; Gschneidner, K.; Harmon, B. Electron correlation effects on the magnetostructural transition and magnetocaloric effect in $Gd_5Si_2Ge_2$. *Phys. Rev. B* **2006**, *73*, 144406.

26. Paudyal, D.; Mudryk, Y.; Pecharsky, V.K.; Gschneidner, K.A. Electronic structure, magnetic properties, and magnetostructural

transition in $Tb_5Si_{2.2}Ge_{1.8}$ from first principles. *Phys. Rev. B* **2011**, *84*, 014421.

27. Pereira, A.M.; Kampert, E.; Moreira, J.M.; Zeitler, U.; Belo, J.H.; Magen, C.; Algarabel, P.A.; Morellon, L.; Ibarra, M.R.; Gonçalves, J.N.; *et al.* Unveiling the (De)coupling of magnetostructural transition nature in magnetocaloric $R_5Si_2Ge_2$(R = Tb, Gd) materials. *Appl. Phys. Lett* **2011**, *99*, 132510.

28. Mudryk, Y.; Paudyal, D.; Pecharsky, V.K.; Gschneidner, K.A.; Misra, S.; Miller, G.J. Controlling Magnetism of a Complex Metallic System Using Atomic Individualism. *Phys. Rev. Lett* **2010**, *105*, 066401.

29. Morellon, L.; Ritter, C.; Magen, C.; Algarabel, P.; Ibarra, M. Magnetic-martensitic transition of $Tb_5Si_2Ge_2$ studied with neutron powder diffraction. *Phys. Rev. B* **2003**, *68*, 024417.

30. Morellon, L.; Blasco, J.; Algarabel, P.A.; Ibarra, M.R. Nature of the first-order antiferromagnetic-ferromagnetic transition in the Ge-rich magnetocaloric compounds $Gd_5(Si_xGe_{1-x})_4$. *Phys. Rev. B* **2000**, *62*, 1022–1026.

31. Levin, E.M.; Pecharsky, V.K.; Gschneidner, K.A. Unusual Magnetic Behavior in $Gd_5(Si_{1.5}Ge_{2.5})$ and $Gd_5(Si_2Ge_2)$. *Phys. Rev. B* **2000**, *62*, 625–628.

32. Casanova, F.; Batlle, X.; Labarta, A.; Marcos, J.; Mañosa, L.; Planes, A. Entropy change and magnetocaloric effect in $Gd_5(Si_xGe_{1-x})_4$. *Phys. Rev. B* **2002**, *66*, 100401.

33. Magen, C.; Algarabel, P.A.; Morellon, L.; Araújo, J.P.; Ritter, C.; Ibarra, M.R.; Pereira, A.M.; Sousa, J.B. Observation of a Griffiths-like Phase in the Magnetocaloric Compound $Tb_5Si_2Ge_2$. *Phys. Rev. Lett* **2006**, *96*, 167201.

34. Arrott, A. Criterion for Ferromagnetism from Observations of Magnetic Isotherms. *Phys. Rev* **1957**, *261*, 2–4.

35. Hadimani, R.L.; Melikhov, Y.; Schlagel, D.L.; Lograsso, T.A.; Jiles, D.C. Study of the Second-Order “Hidden” Phase Transition of the Monoclinic Phase in the Mixed Phase Region of $Gd_5(Si_xGe_{1-x})_4$. *IEEE Trans. Magn* **2012**, *48*, 4070–4073.

36. Hadimani, R.L.; Melikhov, Y.; Snyder, J.E.; Jiles, D.C. Estimation of second order phase transition temperature of the orthorhombic phase of $Gd_5(Si_xGe_{1-x})_4$ using Arrott plots. *J. Appl. Phys* **2008**, *103*, 033906.

37. Pereira, A.M.; dos Santos, A.M.; Magen, C.; Sousa, J.B.; Algarabel, P.A.; Ren, Y.; Ritter, C.; Morellon, L.; Ibarra, M.R.; Araújo, J.P. Understanding the role played by Fe on the tuning of magnetocaloric effect in $Tb_5Si_2Ge_2$. *Appl. Phys. Lett* **2011**, *98*, 122501.

38. Provenzano, V.; Shapiro, A.J.; Shull, R.D. Reduction of hysteresis losses in the magnetic refrigerant $Gd_5Ge_2Si_2$ by the addition of iron. *Nature* **2004**, *429*, 853–857.

39. Carvalho, A.M.G.; Alves, C.S.; de Campos, A.; Coelho, A.A.; Gama, S.; Gandra, F.C.G.; von Ranke, P.J.; Oliveira, N.A. The magnetic and magnetocaloric properties of $Gd_5Ge_2Si_2$ compound under hydrostatic pressure. *J. Appl. Phys* **2005**,*97*.

40. Liu, G.J.; Sun, J.R.; Lin, J.; Xie, Y.W.; Zhao, T.Y.; Zhang, H.W.; Shen, B.G. Entropy changes due to the first-order phase transition in the $Gd_5Si_xGe_{4-x}$ system. *Appl. Phys. Lett* **2006**, *88*, 212505.

41. Gschneidner, K.A.; Mudryk, Y.; Pecharsky, V.K. On the nature of the magnetocaloric effect of the first-order magnetostructural transition. *Scr. Mater* **2012**, *67*, 572–577.

42. Svitelskiy, O.; Suslov, A.; Schlagel, D.; Lograsso, T.; Gschneidner, K.; Pecharsky, V. Elastic properties of $Gd_5Si_2Ge_2$studied with an ultrasonic pulse-echo technique. *Phys. Rev. B* **2006**, *74*, 184105.

43. Pecharsky, A.; Gschneidner, K.; Pecharsky, V.; Schlagel, D.; Lograsso, T. Phase relationships and structural, magnetic, and thermodynamic properties of alloys in the pseudobinary Er_5Si_4-Er_5Ge_4 system. *Phys. Rev. B* **2004**, *70*, 144419.

44. Pecharsky, A.O.; Pecharsky, V.K.; Gschneidner, K.A. Phase relationships and low temperature heat capacities of alloys in the Y_5Si_4–Y_5Ge_4 pseudo binary system. *J. Alloy. Compd* **2004**, *379*, 127–134.

45. Belo, J.H.; Amaral, J.S.; Pereira, A.M.; Amaral, V.S.; Araujo, J.P. On the Curie temperature dependency of the magnetocaloric effect. *Appl. Phys. Lett* **2012**, *242407*, 3–6.

46. Griffiths, R.B. Nonanalytic Behavior Above the Critical Point in a Random Ising Ferromagnet. *Phys. Rev. Lett* **1969**, *23*, 17–19

47. Bray, A. Nature of the Griffiths phase. *Phys. Rev. Lett* **1987**, *59*, 586–589.

48. Zou, J.D.; Liu, J.; Mudryk, Y.; Pecharsky, V.K.; Gschneidner, K.A. Ferromagnetic ordering and Griffiths-like phase behavior in $Gd_5Ge_{3.9}Al_{0.1}$. *J. Appl. Phys* **2013**, *114*, 063904.

49. Ouyang, Z.W. Griffiths-like behavior in Ge-rich magnetocaloric compounds $Gd_5(Si_xGe_{1-x})_4$. *J. Appl. Phys* **2010**, *108*, 033907.

50. Pereira, A.M.; Morellon, L.; Magen, C.; Ventura, J.; Algarabel, P.A.; Ibarra, M.R.; Sousa, J.B.; Araújo, J.P. Griffiths-like phase of magnetocaloric $R_5(Si_xGe_{1-x})_4$ (R = Gd, Tb, Dy, and Ho). *Phys. Rev. B* **2010**, *82*, 172406.

51. Turcaud, J.A.; Pereira, A.M.; Sandeman, K.G.; Amaral, J.S.; Morrison, K.; Berenov, A.; Daoud-Aladine, A.; Cohen, L.F. Spontaneous magnetization above Tc in $La_{0.7}Ca_{0.3}MnO_3$ and $La_{0.7}Ba_{0.3}MnO_3$ polycrystalline materials. *Phys. Rev. B***2014**. in press.

52. Burgy, J.; Mayr, M.; Martin-Mayor, V.; Moreo, A.; Dagotto, E. Colossal Effects in Transition Metal Oxides Caused by Intrinsic Inhomogeneities. *Phys. Rev. Lett* **2001**, *87*, 277202.

53. Oliveira, G.N.P.; Pereira, A.M.; Lopes, A.M.L.; Amaral, J.S.; dos Santos, A.M.; Ren, Y.; Mendonça, T.M.; Sousa, C.T.; Amaral, V.S.; Correia, J.G.; *et al.* Dynamic off-centering of Cr^{3+} ions and short-range magneto-electric clusters in $CdCr_2S_4$. *Phys. Rev. B* **2012**, *86*, 224418.

54. Michelmann, M.; Moshnyaga, V.; Samwer, K. Colossal magnetoelastic effects at the phase transition of $(La_{0.6}Pr_{0.4})_{0.7}Ca_{0.3}MnO_3$. *Phys. Rev. B* **2012**, *85*, 014424.

55. Guo, S.; Young, D.; Macaluso, R.; Browne, D.; Henderson, N.; Chan, J.; Henry, L.; DiTusa, J. Discovery of the Griffiths Phase in the Itinerant Magnetic Semiconductor $Fe_{1-x}Co_xS_2$. *Phys. Rev. Lett* **2008**, *100*, 017209.

9

MICROSTRUCTURE AND SHAPE MEMORY EFFECT OF CU-ZN-NI SHAPE MEMORY ALLOYS

Sathish S[1], U. S. Mallik[2], T. N. Raju[1]

[1]Department of Mechanical Engineering, Dr. Ambedkar Institute of Technology, Bangalore, India

[2]Department of Mechanical Engineering, Siddaganga Institute of Technology, Tumkur, India

ABSTRACT

The microstructure, martensitic transformation behavior and shape memory effect of Cu-Zn-Ni shape memory alloy have been studied by X-ray diffraction (XRD), optical microscopy (OM) and differential scanning calorimetry (DSC). The results show that the recrystallization occurs in the hot-rolled Cu-Zn-Ni alloy by annealing at 800°C and alloy is primarily composed of martensite. A reverse martensite transformation temperature higher than 100°C upon heating has been detected. The alloys exhibit good ductility and shape memory effect (SME). The results obtained are discussed in detail.

KEYWORDS

Cu-Zn-Ni Shape Memory Alloy; Transformation Temperature; X-Ray Diffraction; Optical Microscopy; Shape Memory Effect

1. INTRODUCTION

A shape memory alloy (SMA) is an alloy that remembers its original shape returning the pre-deformed shape upon heating. This material is a lightweight, solid-state alternative to conventional actuators such as hydraulic, pneumatic, and motor-based systems. Shape memory alloys find applications in industries such as medical, robotics and aerospace. Shape memory alloys (SMAs) are metallic materials, which exhibit two unique properties namely, shape memory effect and super-elasticity.

Among many alloy systems which exhibit shape memory effect (SME), Cu-Al-Ni and Cu-Zn-Al shape memory alloys (SMAs) have been studied extensively over the years [1]. Most of the copper-based shape memory alloys are easy to fabricate, process and are less expensive when compared to Ni-Ti shape memory alloys. However, Cu-Al-Ni and Cu-Zn-Al shape memory alloys in the polycrystalline state are brittle and therefore cannot be easily worked due to the high degree of order and high elastic anisotropy of the parent β-phase (austenitic) [2].

Cu-based shape memory alloys are currently derived from three binary alloy systems i.e., Cu-Zn, Cu-Al and Cu-Sn [3]. Among these martensitic transformations, In Cu-Sn alloys are not perfectly thermo elastic and suffer from a rapid degradation of a shape memory properties during ageing at even moderate temperatures. It is also very brittle and the existence of β-composition range is very narrow. These alloys are therefore of theoretical interest than potential commercial shape memory alloys.

Cu-Zn based alloys containing Al, Si, Sn, Ga, or Mn as ternary alloy and Cu-Al based ternary alloys with alloying elements such as Ni, Be, Zn and Mn [4] [5] have been explored for their potential use. Currently Cu-ZnAl and Cu-Al-Ni shape memory alloys are commercially available and put to use in many practical applications. The alloys can also be suitably modified to meet specific requirements for practical applications by selectively alloying with suitable quaternary and grain-refining additions [6]. Accordingly quaternary alloyed SMAs such as Cu-Zn-Al-Ni, Cu-Zn-Al-Ni-Mn and Cu-Zn-Ni-Ti-Mn etc. are developed and put to use in many commercial applications. The alloys are also alloyed with grain refiners such as B, Co, Fe, Ti, V and Zr to reduce the size of grains and improve their mechanical properties [7].

The practical applications of a shape memory alloy depend on its characteristic transformation temperatures and the extent of shape memory effect and superelasticity. Since the shape memory properties of SMAs are sensitive to the composition of the alloy, they play a vital role in determining their transformation temperatures and shape memory characteristics [8]. The main objective of present study is to determine the influence of alloying elements zinc and nickel on the transformation temperatures, morphology of martensite formed, strain recovery by shape memory effect in a broad range of concentrations of Cu-Zn-Ni alloys.

2. EXPERIMENTAL PROCEDURE

From phase diagram [9] pure copper, zinc and nickel were taken in right quantities [Figure 1(a)] to weigh 500 g of the alloy in total and were melted together in an induction furnace under an inert gas atmosphere [10]. The molten alloy was poured into a cast iron mould of dimensions 150 mm × 100 mm × 3 mm and allowed to solidify. The ingots were then homogenized at 800°C in the β-phase for 6 h under an argon atmosphere. The chemical compositions of the alloys were determined using a Perkin-Elmer inductively coupled plasma-optical emission spectrophotometer (ICP-OES), which has the capability to analyze compositions up to an accuracy level of second decimal place [9].

The homogenized alloy samples were hot rolled at 800°C according to phase liquidous diagram from Figure 1(b) to a thickness of 1 mm. The hot rolled samples were homogenized for 5 h at 800°C and step quenched into a boiling water bath (~100°C), followed by quenching them into a water bath at room temperature (~30°C). Step quenching was used to prevent the quench cracks and the pinning of martensitic plates by excess dislocations retained on quenching from high temperature [11] [12]. The shape memory effect is a temperature-dependent effect. During shape memory effect, if the recovery of shape is constrained/hindered, then the alloy generates a force that can be used to do some useful work. The transformation from austenite to martensite and the reverse transformation from martensite to austenite are diffusionless in character and occur over a temperature range. The alloys exhibit four inherent temperatures, which indicate the beginning and end of austenite and martensite transformations. These are known as characteristic transformation temperatures. These temperatures are: M_f—martensite finish; M_s—martensite start; A_s—austenite start; and A_f—austenite finish.

The microstructure and morphology of martensites formed were studied using an optical microscope. The transformation temperatures were determined using a differential scanning calorimeter (DSC) by adopting a heating and cooling rate of 10°C/min. The polycrystalline samples were analyzed by X-ray diffractometer to identify the phases formed at room temperature. The strain recovery by SME of the alloys were determined by bend test on 1 mm thick specimens by measuring the initial and final bend angles just before heating i.e. in the martensitic phase, and immediately after heating, i.e. in the austenitic phase of the alloys [13].

The specimens for microstructural study were prepared by following the standard metallographic practice. The polished specimens were then etched using $K_2Cr_2O_7$ (2 g-$K_2Cr_2O_7$, 8 ml-H_2SO_4, 4 drops-HCl, 100 ml Water) etchant. The etched samples were thoroughly cleaned with acetone and the microstructural examination was carried out using optical microscope to study the morphology and the type of martensite formed.

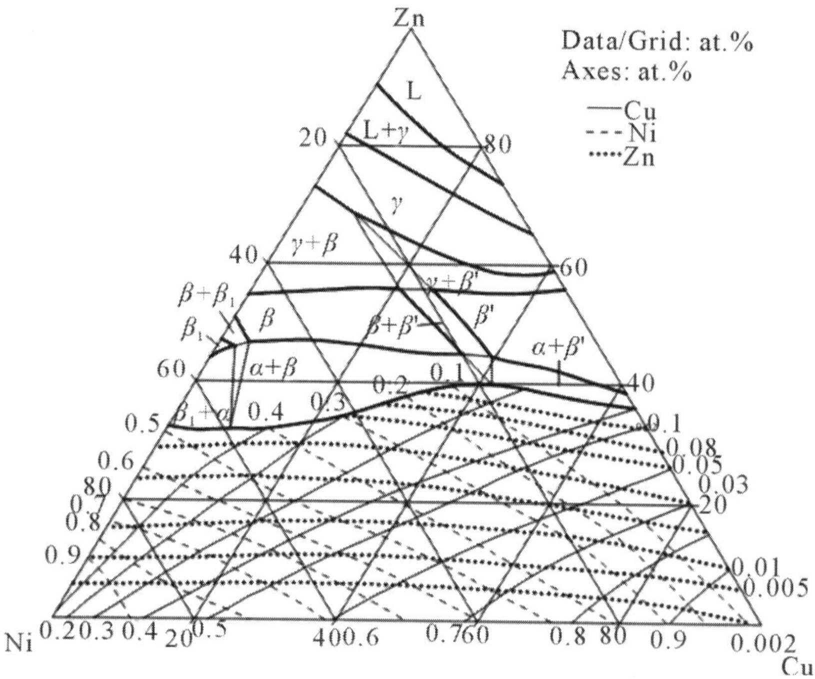

(a)

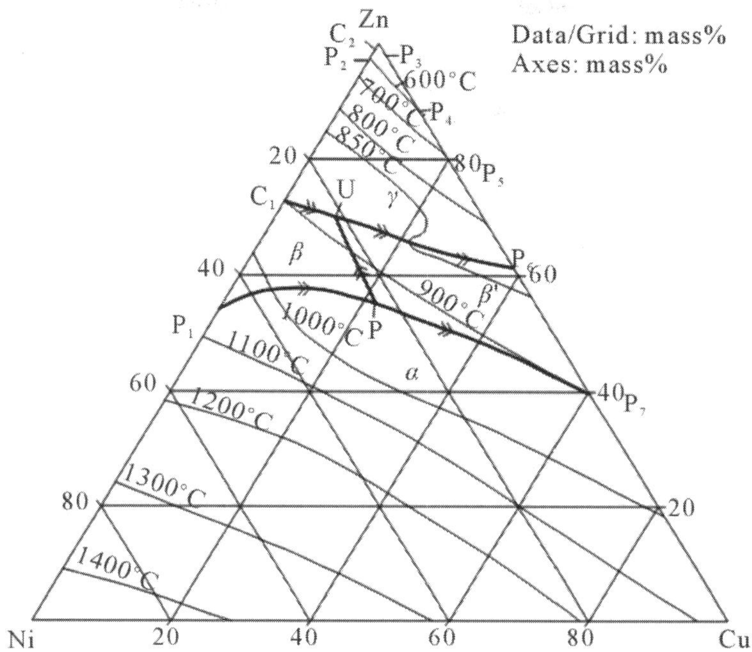

(b)

Figure 1. (a) Isothermal section of Cu-Zn-Ni at 775°C; and (b) Liquidous projections (9).

3. RESULTS AND DISCUSSION

3.1. Chemical Composition, Transformation Temperatures and Microstructures of Ternary Cu-Zn-Ni SMAs

The chemical compositions of the alloys, and the corresponding transformation temperatures, are given in Tables 1 and 2. Table 1 shows the chemical composition and transformation temperatures of the alloys which have almost the same Cu/Ni ratio and varying Zn contents. Table 2 shows the chemical composition and transformation temperatures of the alloys which have almost a constant Cu/Zn ratio and varying Ni contents. During heating martensite transforms to austenite and there is absorption of heat, which is marked by an endothermic peak. On cooling, austenite transforms back to martensite, and heat is released, which is marked by an exothermic peak. The transformation temperature varies with the variation in zinc and nickle contents.

Table 1. Chemical compositions and transformation temperatures of the alloys with almost a constant Cu/Ni ratio and varying Zn contents.

Alloy ID	Chemical Composition (wt. %)			Transformation Temperatures (°C)			
	Cu	Zn	Ni	M_f	M_s	A_s	A_f
1	49.27	45.26	5.47	74	90	95	138
2	48.50	45.59	5.82	82	106	105	145
3	48.22	45.63	6.15	88	112	116	152
4	47.67	46.12	6.21	108	115	118	143

Table 2. Chemical compositions and transformation temperatures of the alloys with almost a constant Cu/Zn ratio and varying Ni contents.

Alloy ID	Chemical Composition (wt%)			Transformation Temperatures (°C)			
	Cu	Zn	Ni	M_f	M_s	A_s	A_f
5	48.96	48.04	2.96	86	111	101	126
6	49.24	46.86	3.75	72	121	116	143
7	49.27	45.91	4.47	66	132	126	156
8	48.50	45.59	5.82	41	143	143	176

From Table 1 it can be observed that as the concentration of zinc increases the transformation temperatures decreases whereas Table 2 it can be observed when the concentration of nickel increases there is increase in the transformation temperatures [14]. From Table 1 and 2, it can also be observed that the martensitic start (M_S) below which the alloy will be completely martensitic and austenitic finish temperature (A_f), above which the alloy is completely austenitic varies depending on the composition of alloys. Therefore these alloys can be suitably designed for use in different temperature and also for low temperature applications [15].

Figure 2 shows the X-ray diffraction patterns of the step quenched alloys. It can be observed fromFigure 2 that there is complete transformation of austenite to martensite when the alloys are step quenched from 800°C to 100°C followed by quenching to room temperature. Depending on the amount of martensite formed the intensities of the peaks corresponding to β_1^1 varies. The amount of martensite formed in turn have a detrimental effect on the shape memory characteristics of the alloys i.e. SME. In all the alloys β_1^1 martensite is predominantly formed. It can also be observed that the alloys with lower Zinc content shows formation of β_1^1 martensite predominantly [Figure 2(a)]. In alloys which have higher contents of Zinc shows formation of γ_1^1 martensite. Invariably in all the alloys there is mixture of both types of martensites.

The optical micrographs of the step quenched alloys are shown in Figure 3. The optical micrograph,Figure 3(a), of the alloy shows the presence of austenite in the as-cast alloy. From Figure 3(b) it can be observed that the alloy with lower zinc content shows formation of lath type β_1^1 martensite with a thin plate like morphology, indicating the complete transformation of austenite to martensite.Figures 3(c) and (d) also show the optical micrographs of the Cu-Zn-Ni alloy samples with higher zinc content, where formation of both kinds of martensites i.e. β_1^1 martensite and γ_1^1 martensite are formed. The γ_1^1 martensite morphology is different from that of β_1^1 martensite. The γ_1^1martensite appears as a spear and also a thick plate-like morphology. In all the alloys it can be observed that there is a complete transformation of the parent austenitic phase to martensite. The different martensitic phases formed in the polycrystalline samples of the step quenched alloys were analyzed using X-ray diffractometers with Cu-Kα radiation, at room temperature.

3.2. Shape Memory Effect (SME)

The alloys were tested for its shape memory effect by using a semi empirical bend test as shown inFigure 4 [6]. The sheet specimens were bent to a U-shape in its martensitic phase by applying an initial strain of 2% around a round bar of diameter 50 mm. The alloys were then heated to above their austenitic finish temperature. The corresponding angles of bend before heating and after heating to above their austenitic finish temperature were measured. The percentage of strain recovery by shape memory effect was determined by using the relation:

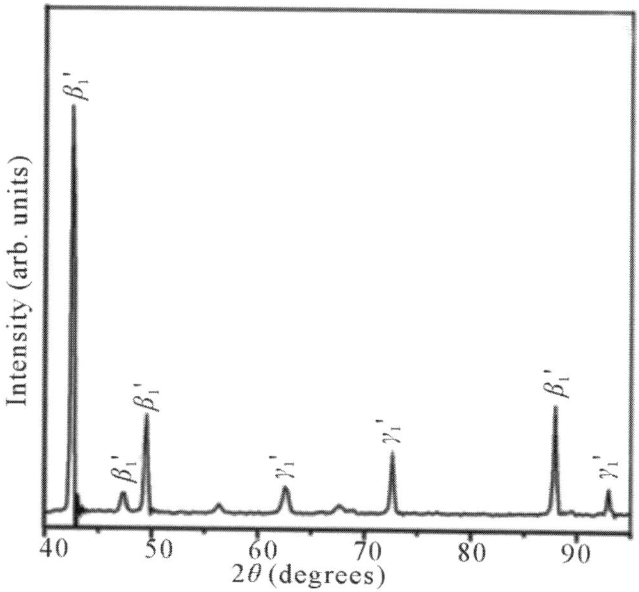

(a)

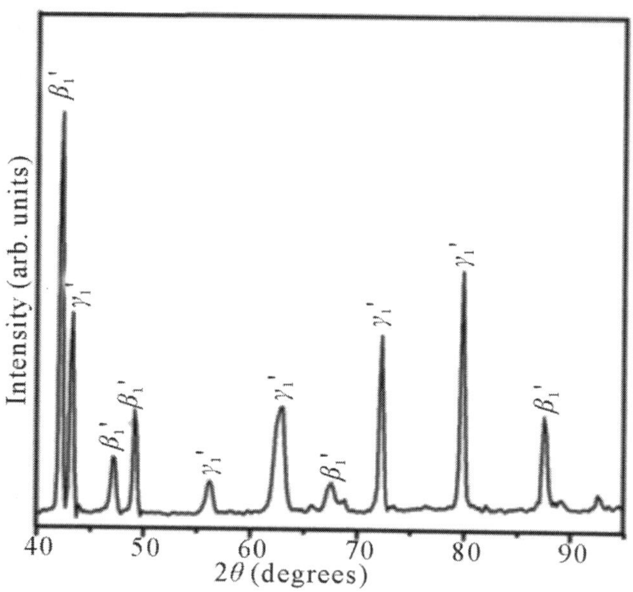

(b)

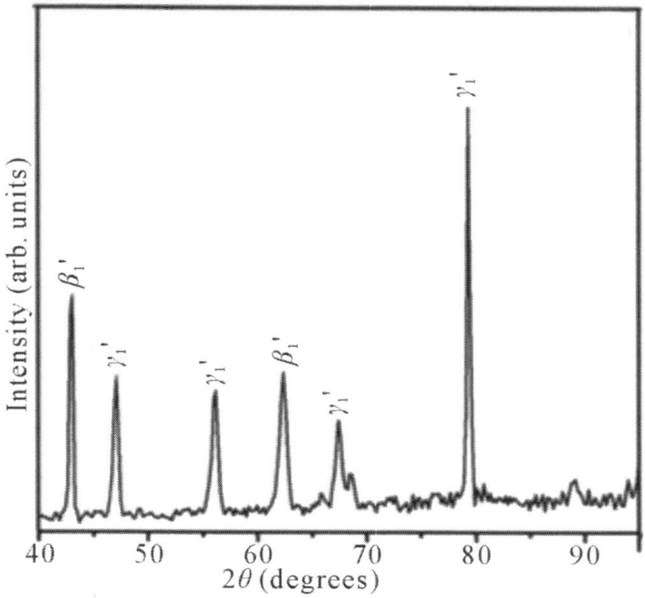

(c)

Figure 2. X-ray diffraction patterns of Cu-Zn-Ni SMAs. (a) Alloy 2; (b) Alloy 4; and (c) Alloy 5.

$$SME = \frac{\theta_m}{180 - \theta_e};$$

$$\varepsilon = \frac{t}{d}$$

where θ_e = angle recovered on unloading, θ_m = angle recovered on heating, d = 50 mm, t = 1 mm and e = 2%.

In the martensitic phase of the alloys at T < Martensite finish (M_f). The alloys were then heated to above their Austenite finish (A_f) temperature. The corresponding angles of bend before heating and after heating to above A_f temperature were measured and the percentage of strain recovery was determined. These alloy system exhibit significant strain recovery by SME to the extent of 70% to 99% as shown in Table

3. The amount of shape memory depends mainly on the amount of martensite present in the alloy and the extent of transformation of martensite to austenite.

Figure 3. Optical micrographs of Cu-Zn-Ni SMAs exhibiting lath type and spear type martensite structure obtained on step quenching. (a) Alloy 2; (b) Alloy 4; (c) Alloy 5; and (d) Alloy 7.

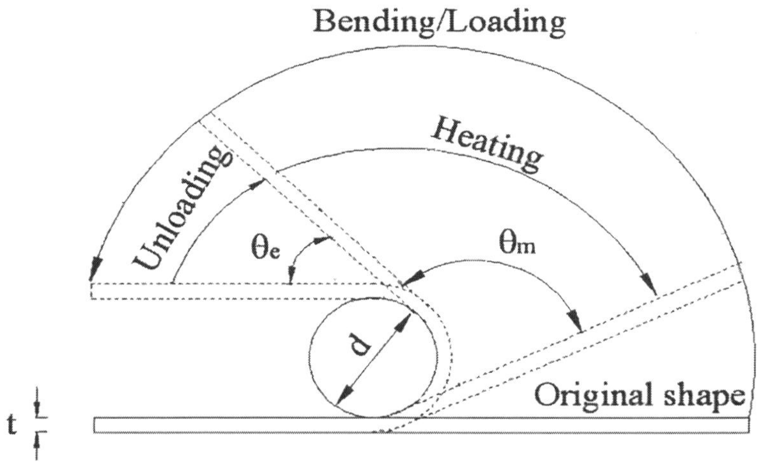

Figure 4. Schematic diagram of the bend test to determine strain recovery by SME.

If there is any residual martensite remaining in the austenite phase of the alloy the strain recovery will be reduced proportionately [16]. The other factors which influence the strain recovery by SME are the imperfections in the material, such as quenched in vacancies, dislocations, etc., which cause the pinning of martensitic plates, thereby hampering the transformation and in turn the strain recovery. These can be reasonably avoided by step quenching. But it has been observed, that the alloys with the higher concentration of Zinc and lesser concentration of nickel exhibit higher recovery strain by SME. It may be attributed to the ease with which the martensitic to austenitic transformation occurs, and the formation of a higher proportion γ_1' martensite compared with that of β_1' martensite in the alloys. This also leads to decrease in the transformation temperatures.

4. CONCLUSIONS

1) The type and amount of martensite formed in these alloys are mainly dependent upon the amount of Zinc and Nickel in the alloys. β_1' martensitic phase is predominant in the composition range of the alloys chosen. Two kinds of martensites are observed in these alloys i.e. β_1^1 martensite and γ_1' martensite with high density of twins. The β_1^1 martensite is formed when the content of zinc is less and γ_1' martensite is formed when the content of zinc is more in the composition range of the studied alloys.

2) The transformation temperatures are highly sensitive to the variation in zinc concentrations of the alloy. As the content of zinc increases, the transformation temperatures decrease, whereas as the content of nickel increases, the transformation temperatures increase.

3) The alloys exhibit good strain recovery by SME. A strain recovery of up to 99% by SME was observed inthese alloys. The variation in zinc and nickel in the alloys concentration does not significantly affect the strain recovery by SME but depends on the amount and type of martensite transformed.

Table 3. Strain recovery by SME of Cu-Zn-Ni SMAs.

Alloy ID	Composition (wt%)			Strain Recovery (%)
	Cu	Zn	Ni	
2	48.50	45.59	5.82	70
4	47.67	46.12	6.21	85
5	48.96	48.04	2.96	99
7	49.27	45.91	4.47	87

REFERENCES

1. Duerig, T.W., Melton, K.N., Stockel, D. and Wayman, C.M. (1990) Engineering Aspects of Shape Memory Alloys. Butterworth Heinemann Publishing, London, 3-20.

2. Sutou, Y., Omori, T., Kainuma, R., Ono, N. and Ishida, K. (2002) Enhancement of Superelasticity in Cu-Al-Mn-Ni Shape Memory Alloys by Texture Control. Metallurgical and Materials Transactions A, 33A, 2817-2824.

3. Ahlers, M. (2008) The Martensitic Transformation in the Cu-Zn Based Shape Memory Alloys as a Tool for the Evaluation of Transformation Mechanisms and Phase Stabilities. Materials Science and Engineering A, 481-482, 500-503.

4. Kainuma, R., Takahashi, S. and Ishida, K. (1996) Thermoelastic Martensite and Shape Memory Effect in Ductile Cu-Al-Mn Alloys. Metallurgical and Materials Transactions A, 27A, 2187-2195.

5. Kainuma, R., Satoh, N., Liu, X.J., Ohnuma, I. and Ishida, K. (1998) Phase Equilibria and Heusler Phase Stability in Cu-Rich Portion of the Cu-Al-Mn System. Journal of Alloys and Compounds, 266, 191-200.

6. Mallik U.S. and Sampath, V. (2008) Influence of Aluminum and Manganese Concentration on the Shape Memory Characteristics of Cu-Al-Mn Shape Memory Alloys. Journal of Alloys and Compounds, 459, 142-147.

7. Mallik, U.S. and Sampath, V. (2008) Effect of Composition and Ageing on Damping Characteristics of Cu-Al-Mn Shape Memory Alloys. Materials Science and Engineering A, 478, 48-55.

8. Sutou, Y., Kainuma, R. and Ishida, K. (1999) Effect of Alloying Elements on the Shape Memory Properties of Ductile Cu-Al-Mn Alloys. Materials Science and Engineering A, 273, 375-379.

9. ASM (1992) ASM Handbook Volume 03: Alloy Phase Diagrams. 10th Edition, ASM International, Almere.

10. Mallik, U.S. and Sampath, V. (2006) Influence of Composition on Shape Memory Characteristics of Cu-Al-Mn Shape Memory Alloys. Proceedings of International Conference on Advances in Materials and Materials Processing (ICAMMP-2006), Kharagpur, 3-5 February 2006, 583-588.

11. Lopez del Castillo, C., Blazquez, M.L., Gomez, C., Mellor, B.G., de Diego, N. and del Rio, J. (1988) The Stabilization of Martensite in Cu-Al-Mn Alloys. Journal of Materials Science, 23, 3379-3382.

12. Husain, S.W. and Clapp, P.C. (1988) The Intergranular Embrittlement of Cu-Al-Ni β-Phase Alloys. Journal of Materials Science, 22, 2351-2356.

13. Kim, H.W. (2004) A Study of the Two-Way Shape Memory Effect in Cu-Zn-Al Alloys by the Thermomechanical Cycling Method. Journal of Materials Processing Technology, 146, 326-329.

14. Sarı, T.K. (2008) Effects of Deformation on Microstructure and Mechanical Properties of a Cu-Al-Ni Shape Memory Alloy. Material Characterizations, 59, 920-929.

15. Shi, D., Gui, J., Tan, S.S. and Wang, R. (1998) Effect of Ageing in the Two-Phase Region in a Cu-Zn-Al Shape Memory Alloy. Materials Science and Engineering B, 56, 31-36.

16. Morris, M.A. and Lipe, T. (1994) Microstructural Influence of Mn Additions on Thermoelastic and Pseudoelastic Properties of Cu-Al-Ni Alloys. Acta Metallurgica et Materialia, 42, 1583-1594.

10

SMART CYMBAL TRANSDUCERS WITH NITINOL END CAPS TUNABLE TO MULTIPLE OPERATING FREQUENCIES

Andrew Feeney and Margaret Lucas

School of Engineering, University of Glasgow, Glasgow UK

ABSTRACT

Cymbal flextensional transducers have principally been adopted for sensing and actuation and their performance in higher power applications has only recently been investigated. Nitinol is a shape-memory alloy (SMA) with excellent strain recovery, durability, corrosion resistance, and fatigue strength. Although it has been incorporated in many applications, the implementation of nitinol, or any of the SMAs, in power ultrasonic applications is limited. Nitinol exhibits two phenomena, the first being the superelastic effect and the second being the shape-memory effect (SME). This paper assesses two cymbal transducers, one assembled with superelastic nitinol end caps and the other with shape-memory nitinol end caps. Characterization of the nitinol alloy before the design of such transducers is vital, so that they can be tuned to the desired operating frequencies. It is shown this can be achieved for shape-memory nitinol using differential scanning calorimetry (DSC); however, it is also shown that characterizing superelastic nitinol with DSC is problematic. Two transducers are assembled whose two operating frequencies can be tuned, and their dynamic behaviors are compared. Both transducers are shown to be tunable, with limitation for high-power applications largely being associated with the bond layer.

SECTION I

INTRODUCTION

CYMBAL transducers were developed in the 1990s [1], and have been widely adopted in low-power ultrasonic applications, such as in hydrophones and for sensing and actuation, because of their small size, capability of delivering relatively high output power, and simplicity of design and assembly. However, there are also benefits for their incorporation in high-power ultrasonic devices. Cymbals offer an opportunity to design devices where there is less need, compared with devices based on a Langevin transducer, to restrict the attached tool geometry to resonant and high-gain configurations. The development of cymbal transducers that can be tuned to more than one frequency, by integrating shape-memory alloys, is particularly applicable to many ultrasonic surgical devices, for which tailoring the performance for different hard and soft tissues can offer significant procedure enhancements.

The cymbal transducer, a schematic of which is shown in Fig. 1, consists of a piezoceramic disc sandwiched between two metal end caps; the low-displacement radial motion of the disc causes a high-displacement axial motion of the end caps [2].The size of a cymbal transducer is typically around 12 mm in diameter when used in ultrasound systems such as underwater hydrophone arrays [3]. The resonant frequency and axial displacement amplitudes are dependent on the geometrical dimensions of the cavity and the elastic properties of the end-cap material [4], with metals such as brass, steel, and titanium being commonly adopted [5], [6] primarily because they possess desirable elastic properties for sonar and hydrophone applications. In addition, the ease with which the end caps can be manufactured means that the geometrical dimensions of the end cap cavities can be readily tuned to generate a desired resonant frequency response. The material and geometry determine the displacement amplitude that this type of device can achieve. For example, titanium is stiffer than brass, and so will result in a device with a higher resonant frequency, but lower displacement amplitude. It has been found that, in general, cymbal transducers offer the advantage of being straightforward and inexpensive to manufacture and assemble [7] and this is especially true when end caps can be punched from a sheet metal, such as for brass, steel, and titanium.

A. Cymbal Transducer Applications

There have been many adaptations of the classical design of the cymbal transducer for different applications, although these have been largely for developments in sonar and hydrophone devices [8], which are low to medium power. The cymbal transducer is particularly suited to underwater applications because the device itself is small compared with the wavelength at resonance in water [2]. Individually, a cymbal transducer is relatively inefficient for hydrophone applications, albeit with a desirable Q-factor. Examples of improvements and design changes include a cymbal transducer with concave end caps developed to operate at high hydrostatic pressure [9], and cymbals set in array formations to enable greater radiation efficiency, output power, and directivity compared with single element formations [8], [10]. Array configurations have also been used for transdermal drug delivery via sonophoresis [11], [12]. Cymbals have been developed as miniaturized actuators [13], as transducers for applications in energy harvesting [14], and have recently been studied for highpower ultrasonic applications [15], [16], leading to modifications of its physical coupling mechanisms to allow for excitation of the associated higher displacements [16].

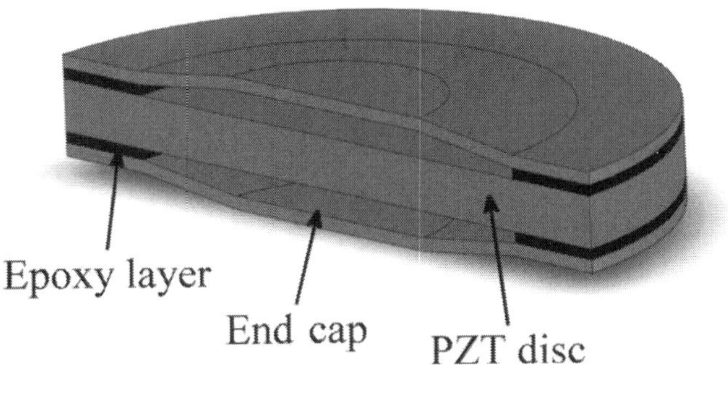

Figure 1. Cut-away diagram of a cymbal transducer.

Research has been conducted on the characterization of a power ultrasonic cymbal transducer configuration [15], and it is clear that in many applications there is a need for devices to operate at more than one frequency. For example, selectable-frequency ultrasonic cutting or drilling devices can allow improved penetration through layers of different materials, and cavitation activity in high-power ultrasonic vessels can be controlled by using selectable excitation frequencies to generate more tailored acoustic pressure fields. This study focuses on a modification of the design to develop a device which exhibits a tunable frequency capability. One transducer configuration that has the potential to deliver this is an adaptation of a cymbal transducer that incorporates a shape-memory alloy (SMA).

B. The Active Material Nitinol

Nickel-titanium (Ni-Ti) is a shape-memory alloy which possesses properties that were first recognized at the Naval Ordnance Laboratory (a former US Naval laboratory) in the 1960s, where it was eponymously referred to as nitinol, after its place of discovery [17]. Ni-Tibased SMAs possess a high level of strain recovery, exhibit good corrosion resistance resulting from a passive titanium-rich oxide layer which can form on the surface of the material, can generally endure high stress levels, and are biocompatible [18], [19]. Nitinol has therefore been employed in products such as biomedical stents, orthodontic archwires, frames for spectacles, and minimally-invasive surgical devices [20]. Nitinol drills have been developed which exploit the capability of the material to deform through large angles but still perform as required. Superelastic nitinol has also been found to be attractive for the vibration control of large structures because of its high energy-dissipation capability.

Nitinol is a binary alloy which undergoes phase transitions between high-temperature austenite and low-temperature martensite, and these phases have different elastic properties and crystal structure. The elastic modulus can typically be between 20 and 40 GPa in the martensite phase and 60 to 80 GPa in the austenite phase. Forward and reverse phase transformations between cubic B2 austenite and monoclinic B19' martensite, occur in nitinol under the application of stress or a change in temperature. However a third phase is often exhibited in nitinol alloys, especially those which are nickel-rich and have been cold worked or subjected to certain aging heat treatments, for example, within the 350°C to 500°C temperature range. This third phase is called the R-phase, which is a rhombohedral alteration of B2 austenite [19]. It is common for the R-phase to be misinterpreted as martensite, but can usually be distinguished from martensite because there is normally a very low thermal hysteresis between austenite and the R-phase, as low as 2°C [19], much smaller than the austenite-martensite hysteresis which has poorer cyclic repeatability [19]. The phase transformations in nitinol occur via shear lattice distortion [20] and there are normally six characteristic transformation temperatures which can be identified for a particular nitinol alloy. These are the start and finish of the austenite (A), rhombohedral (R), and martensite (M) phases. The rhombohedral R-phase represents an intermediate phase which can appear when cooling from austenite to martensite; however, in several cases the R-phase can also intermediately appear in the transition from martensite to austenite [19], [21], with the corresponding transformation temperatures being designated as and [21]. A particular phase has not fully formed until the respective finish temperature has been passed. As an example, if a nitinol alloy is transforming from R-phase to austenite, there will exist a mixture of R-phase and austenite in between the and A_F transformation temperatures. This will be observed until the A_F transformation temperature is reached, when the alloy microstructure has fully oriented to the austenite phase.

The properties which nitinol exhibits that are relevant and important to their adoption in tunable frequency high-power ultrasonic transducers are superelasticity and the shape-memory effect (SME). The superelastic effect is an isothermal process which occurs above the material's final austenitic transformation temperature (A_F) when it experiences a stress causing it to transform to martensite [19]. This enables a higher deformation to be recovered. The transformation temperatures are dependent on stress [17], [20], such that the material undergoes a change in microstructure, reorienting the austenite phase to the martensite phase. This effect will reverse upon unloading, because of the instability of the martensite phase at temperatures at which austenite exists. In addition, the thermal response, as with the stress-strain response, is hysteretic [22].

The shape-memory effect is stimulated via thermal loading of the material, in which a phase transition is produced by heating or cooling the nitinol to a specific characteristic transformation temperature [20]. One interesting aspect of this phenomenon is that if a nitinol component has a particular shape when it exists as martensite and is then deformed, apparently plastically, the component can be heated above its final austenitic transformation temperature, when it will recover its original (pre-deformation) shape, hence the term shape memory. Upon cooling, the component will retain its original shape. If the component is set to the same shape for both martensite and austenite phases, then the thermal loading will transform the microstructure from one phase to the other with no shape change.

There have been recent studies which have focused on the martensitic transformations in nitinol as a result of ultrasonic vibrations [23], [24]. It has been found that phase transformations can be induced by the application of ultrasonic energy in a nitinol alloy to produce the SME [23], and that application of ultrasonic vibrations which cause the nitinol specimen to experience cyclic stress will result in a reduced A_F. The implementation of SMAs in ultrasonic devices therefore must be based on an understanding of the effect of ultrasonic vibrations on the phase transformation behavior of the material, especially the transformation temperatures.

C. Tunable and Multiple Frequency Devices

There have been several studies reporting the development of transducers which can operate in different modes of vibration. One study has demonstrated that cymbal transducers fabricated using nitinol end caps can be used to tune the resonant frequency, and have the ability of active repair, using the shape-memory effect, for damage incurred at high hydrostatic pressures, at which the end caps can collapse [25]. Other successful developments in tunable and multiple frequency devices include a tunable vibration absorber which incorporates the magnetostrictive material Terfenol-D to control the resonant frequency of the absorber via a change of the elastic properties as a result of the magnetostrictive effect [26], a tunable ultrasonic dental scaler [27] which uses magnetostrictive scaler inserts to change the resonant frequency of the device, and a tunable ultrasonic cutting device [28] that can change resonant frequency by adjustment of the length of specific components. Nitinol has been used in tunable devices which are not ultrasonic, for example as adaptive tuned vibration absorbers (ATVAs) [29]. Nitinol has been exploited successfully in a wide range of applications, but there has been only limited research focusing on the integration of nitinol in ultrasonic transducers, especially in the development of multiple frequency or tunable devices.

The vibration behavior of two ultrasonic cymbal transducers is discussed here. The transducers are fabricated using two different end cap nitinol alloys, superelastic ni-tinol end caps (Johnson Matthey Noble Metals, London, UK) and shape-memory nitinol end caps (Memry GmbH, Weil am Rhein, Germany). The results show how tunable devices incorporating nitinol end caps can be successfully created and characterized to operate at two selectable frequencies. However, characterizing the transformation temperatures of superelastic and shape-memory nitinol at the device design stage, to tailor transducers to operate at specific frequencies, relies on measurements that are often misinterpreted and can be inaccurate. These measurements and their implications for design of tunable devices are also discussed.

SECTION II

TRANSDUCER MANUFACTURE ANDASSEMBLY

Commonly, cymbal transducer end caps are punched from sheet metal to ensure accuracy and repeatability in the geometry of the end cap cavities. However, there are special considerations associated with the fabrication of nitinol end caps because cold-working, chemical composition, and heat treatments can all influence the transformation temperatures. The superelastic nitinol end caps which were used in this study were manufactured to be in an oxide-free superelastic condition, with the fabrication process utilizing shape-set annealing. The binary alloy was composed of 55.99wt% Ni with a balance of Ti. Small quantities of C, O*Co*, Cr, Cu, Fe, Mn, Mo, Nb, Si, and W, making up under 0.3% of the total alloy composition, were used in conjunction with precise cold-working and heat treatments to fine-tune the transformation temperatures of the material. This ensures the material is superelastic at room temperature. The shape-memory nitinol end caps were produced using cold rolling to have a final austenitic transformation temperature of $60 \pm 10°C$. The shape-memory binary alloy was composed of 55.11wt% Ni with a balance of Ti and small quantities of Cr, Cu, Fe, Nb, *Co*, C, Oand H, and was fabricated according to ASTM F2063–05 [30]. Again, precise cold-working and heat treatments were implemented to fine-tune the transformation temperatures of the material. It is difficult to manufacture end caps to the precise dimensions desired, and small variations in dimensions can have a significant influence on the resonant frequency response [4]. This is complicated by the fact that the transformation properties of the nitinol must be retained, and so the final component is produced using a careful balance between heat treatments, cold-working, and chemical composition.

Two transducers incorporating end caps of the two materials, superelastic nitinol and shape-memory nitinol, were fabricated according to the data shown in Table I. A hard PZT disc in the form of Navy Type-I Sonox P4 (CeramTec, Plochingen, Germany) constituted the piezo ceramic driver material for each transducer. A layer of insulating epoxy resin (Eccobond, Ellsworth Adhesives Ltd., East Kilbride, UK) at a ratio of three parts epoxy resin to one part resin hardener was applied to the flange of each end cap before assembly of the transducer, which was left to cure at room temperature for 24 h.Fig. 2 shows a fully-assembled nitinol cymbal transducer.

The epoxy layer is nominally 40 thick, but is often difficult to precisely control because of the viscous nature of the deposited adhesive. In this study, an insulating, nonconductive epoxy was chosen because conductive epoxies commonly have lower bond strength [2]. Any inhomogeneities in the epoxy bond layer are usually a consequence of roughness of the end cap flange surfaces, air bubbles, and nonuniformity of deposition, which can all contribute to variations in resonant frequency [31]. Therefore, during the curing process, it was ensured that pressure was only applied to the flange of each end cap via the use of a custom rig, eliminating as many defects in the bond layer as possible and also ensuring the uniform application of pressure around each flange circumference. To ensure electrical contact between the end cap flanges and the piezoceramic disc, a small solder spot was applied to each surface of the PZT. The end caps were then bonded to the piezoceramic disc, before thin electrode wires were bonded to the flange surface of each end cap with a conductive silver epoxy (Chemtronics, Kennesaw, GA). Using a conductive epoxy to bond the electrode wires to the transducer end cap flange surfaces was preferred because of several known issues surrounding the soldering of materials to nitinol. It is known that nitinol can be difficult to solder because of the passive Ti-rich oxide layer which can form on the surface; also, any heating of the material could have undesirable effects on the material transformation temperatures [18].

Table I. Cymbal transducer dimensions (in Millimeters).

Transducer	Cavity apex diameter	Cavity base diameter	Cavity depth	End cap thickness	PZT thickness	Transducer diameter
Superelastic nitinol end caps	4.50	9.10	0.31	0.24	1.00	12.72
Shape-memory nitinol end caps	4.50	9.70	0.26	0.24	1.00	12.69

SECTION III

PHASE TRANSFORMATIONCHARACTERIZATION

There were four steps employed to characterize the transducers, comprising differential scanning calorimetry (DSC), electrical impedance measurements (4294A, Agilent Technologies Inc., Santa Clara, CA), experimental modal analysis (EMA) using a 3-D laser Doppler vibrometer (LDV) (CLV, Polytec GmbH, Waldbronn, Germany) and vibration resonance response characterization using a 1-D LDV (Polytec CLV) to observe the displacement amplitude of each transducer at selected input voltages. DSC (Diamond DSC, Perkin Elmer, Waltham, MA; calibrated using indium) was first conducted, at scan rates of *10°C/*min and 20°C/min, to obtain the transformation temperatures of the material; Fig. 3 shows the DSC analysis results for the nitinol alloys used in this study.

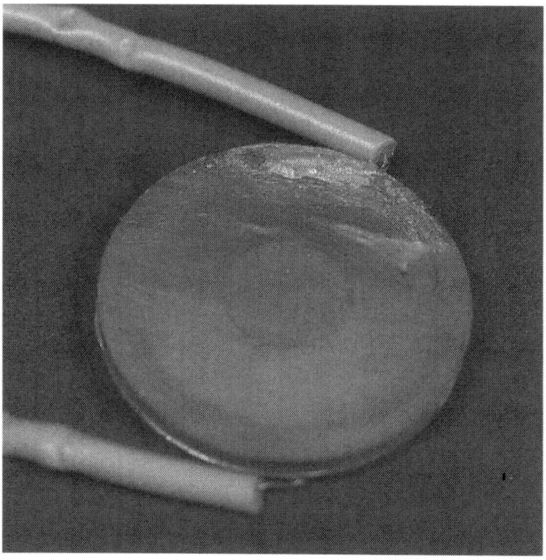

Figure 2. A nitinol cymbal transducer.

A sample of each nitinol alloy is exposed to a controlled thermal cycle by being heated to a predetermined temperature before being cooled again. DSC analysis allows the transformation temperatures of the material to be identified, where the endothermic reaction on heating signifies a transition to austenite, and the exothermic reaction on the cooling part of the cycle signifies a transformation to martensite. Heat flow in the sample is presented as a function of temperature. The heat flow detected in a cycle increases as the scan rate is raised because the material is being subjected to the same temperature range but over a shorter period of time. The DSC contains two furnaces, one for the sample material and the other for a reference. The power which is supplied to both furnaces is measured and monitored by the DSC, and so the increase in heat flow requires an increase in power. There are differences in the results for each alloy between the two scan rates because of this increased sensitivity to heat flow changes. The result is more discernible peaks and troughs in the response for the higher scan rate, but reduced accuracy of transformation temperature data because of the consequential thermal lag. A compromise must therefore be made in DSC analysis.

Nitinol exhibits thermal hysteresis, and this phenomenon can be observed in Fig. 3. This means that the heating and cooling cycles differ, resulting in several transformation temperatures for one particular alloy which all correspond to the start and finish of the respective material phases. Consequently, the temperature ranges for which each phase appears can be broadly identified, with the start and finish transformation temperatures extracted from the data by the intersection of the steepest slopes of the curves with the chosen baseline [19]. The transformation temperature values which are identified from the traces are therefore approximations, because there are errors associated with interpretation of these intersection points [19]. This is a limitation of the DSC technique, but in general the phase transition for each alloy is observed when the temperature is between A_S and A_F. The resulting transformation temperature information from the DSC analysis is presented in Table II.

The data in Table II shows that for both alloys the transformation temperatures for a scan rate of 20°C/min are marginally higher than those identified for 10°C/min, as expected. The superelastic alloy results, shown in Figs. 3(a) and 3(b) for scan rates of 10°C/min and 20°C/min respectively, show that a very small thermal hysteresis has been detected between the austenite phase and the phase observed on the cooling cycle.This phase is the R-phase, identifiable because the thermal hysteresis between the austenite phase and that which was detected on the cooling cycle is too low for it to be a martensite phase, the thermal hysteresis being much greater for a classical austenite to martensite transition [19]. The martensite phase is not observed as it was outside the range of the test setup. The DSC results indicate that the superelastic material exhibits superelasticity from a temperature of approximately *25°C*, at A_F. This means that above this temperature, sufficient stress will cause the material to reorient from austenite to stress-induced martensite, thus changing the elastic properties of the alloy. For a cymbal transducer with superelastic nitinol end caps, this should be exhibited as a change in the vibration response.

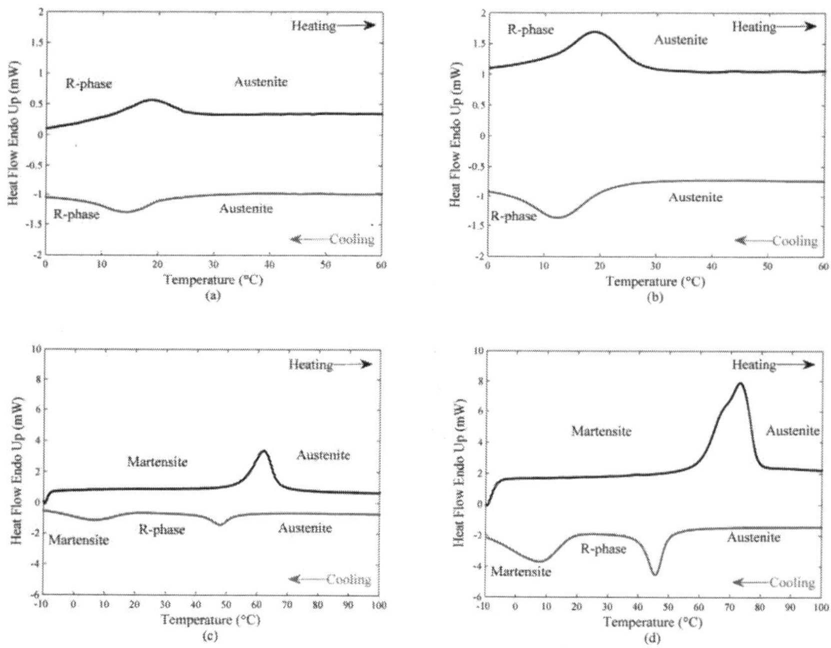

Figure. 3. DSC of superelastic nitinol end-cap material at (a) 10°c/min and (b) 20°c/min, and shape-memory nitinol end-cap material at (c) 10°c/min and (d) 20°c/min.

An intermediate R-phase is also observed in the cooling cycle of the shape-memory material, as can be observed from Figs. 3(c) and 3(d), and it is clear that the shape-memory material exists as austenite in the region above approximately *55°C*. There is also evidence of the emergence of an intermediate R-phase in the heating cycle of the shape-memory alloy at a scan rate of 20°C/min [Fig. 3(d)], with the R-phase peak overlapped with the austenite peak, which can often occur [19]. This was not observed in the lower scan rate of 10°C/min [Fig. 3(c)], highlighting the advantage of increased sensitivity in the data. For transducer design, it is important to identify all possible phase changes in the temperature range within which the device is operating to avoid unwanted frequency changes. It is therefore advisable to characterize the material at more than one scan rate, a lower scan rate for transformation temperature accuracy and a higher scan rate for improved sensitivity, to have a clearer understanding of the transformation behavior.

Table II. Nitinol transformation temperatures (°C) extracted from DSC analysis.

	10°C/min						20°C/min					
Scan rate	A_F	A_S	M_F	M_S	R_F	R_S	A_F	A_S	M_F	M_S	R_F	R_S
Superelastic	23	16	—	—	9	19	25	15	—	—	7	19
Shape memory	68	55	−6	16	44	52	78	61	−5	17	40	50

The DSC results for the alloys show that the material can exist as either martensite or R-phase at room temperature. It is assumed that both materials exist as the R-phase for the following reasons. For the shape-memory alloy, the nitinol is fabricated at high temperatures [19], requiring cooling to produce the final condition, and the material is therefore delivered from a cooling cycle, consistent with existence in the R-phase. For this material, this was checked by conducting the DSC over a larger temperature range, cooling until the martensite phase was observed in the thermograms. If the material was supplied in the R-phase form for both alloy types, this means the microstructure should remain as R-phase when heated from room temperature until it transforms to austenite. This has significance for the design of tunable transducers. Instead of utilizing the phase change between austenite and martensite (as is usual for SMA devices) where there is a large thermal hysteresis, a smaller temperature difference can be utilized to generate a comparable frequency shift, by taking advantage of the austenite—to—R-phase transformation. This provides an opportunity to design a cymbal transducer for which two selectable resonant frequencies can be achieved through a very small change in temperature, possibly as low as 5°C.

Table III. Resonant frequencies (In HERTZ) of the 2 cymbal transducers.

Superelastic nitinol			Shape-memory nitinol			
R-phase (22°C)	Austenite (44°C)	Change	R-phase (22°C)	Austenite (74°C)	Change	Mode
23 495	28 137	+4642 (+19.76%)	23 525	22 775	−750 (−3.19%)	Asymmetric
32 698	33 407	+709 (+2.17%)	25 650	28 800	+3150 (+12.28%)	Symmetric

Electrical impedance measurements were made to verify the quality of the assembled transducers, as well as to monitor the resonant frequencies and note the temperatures at which these changed. Impedance analysis is conducted at a very low excitation voltage (0.5 which imposes minimal stress on the nitinol, so it can be expected that only temperature effects on the impedance response are measured. The cymbal transducer under test was placed in an environmental chamber where temperature was controlled and monitored using an IR thermal camera (FLIR T425, FLIR Systems, Wilsonville, OR), to monitor the temperature of the cymbal end cap, in conjunction with a type-K thermocouple, which was used to monitor the chamber temperature. The impedance frequency spectra of the two transducers in both the R-phase and austenite phase were measured (Fig. 4) and the resonant frequencies are recorded in Table III.

Fig. 4 and Table III demonstrate that the resonant frequencies of both transducers have been altered by a phase change. The troughs in the impedance—frequency spectra shown in Fig. 4 occur at the resonant frequencies associated with the axial modes of the transducers and the frequency shifts between the two measurement temperatures is clear. It is also noticeable for each transducer that there are two resonance responses in the impedance—frequency spectra, these being associated with two modes, the symmetric mode and asymmetric mode. The frequency of the asymmetric mode does not change as much as that of the symmetric mode between the two measurement temperatures.

The DSC results of the superelastic nitinol alloy do not correlate well with the transformation temperature required to shift the resonant frequency of the superelastic nitinol cymbal transducer. The impedance—frequency spectrum was continuously monitored as the transducer was heated to observe in real time the effect of any phase change on the response of the transducer, and it was clearly observed that the transformation to the high-temperature austenite phase occurred close to 40°C. This exhibited as a sudden increase in resonant frequency in the impedance—frequency trace during the experiment. This phase change was recorded at a temperature approximately 20°C higher than the DSC analysis suggested. For this reason, the impedance analysis was recorded at a temperature of 44°C, as shown in Fig. 4(a), to ensure the measured response of the transducer was in the austenite phase. In contrast, the shape-memory nitinol transducer transformed at temperatures consistent with the DSC analysis data, resulting in the resonant frequency shift shown in Fig. 4(b).

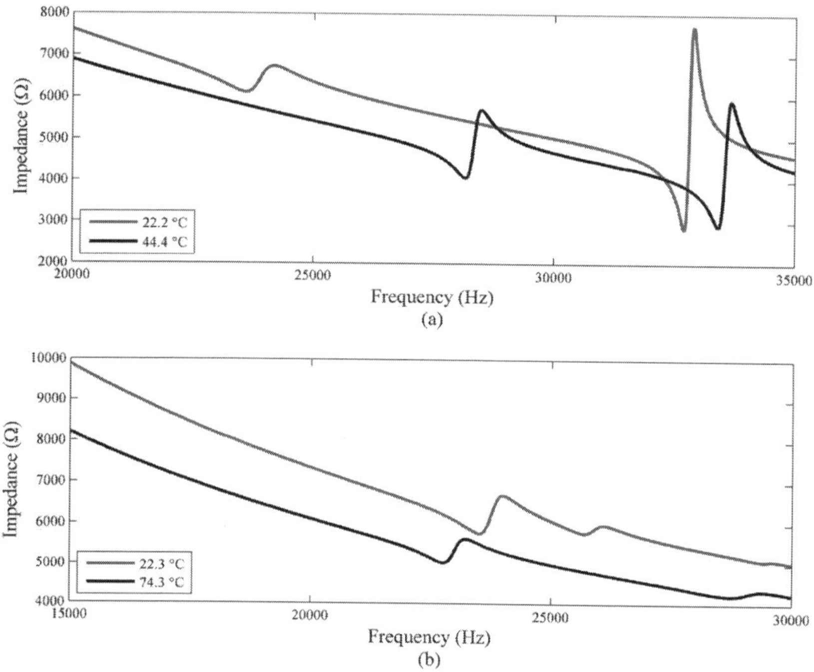

Figure 4. Impedance-frequency spectra of the symmetric and asymmetric mode resonant frequencies for the (a) superelastic and (b) shape-memory transducer at two temperatures associated with a phase transition.

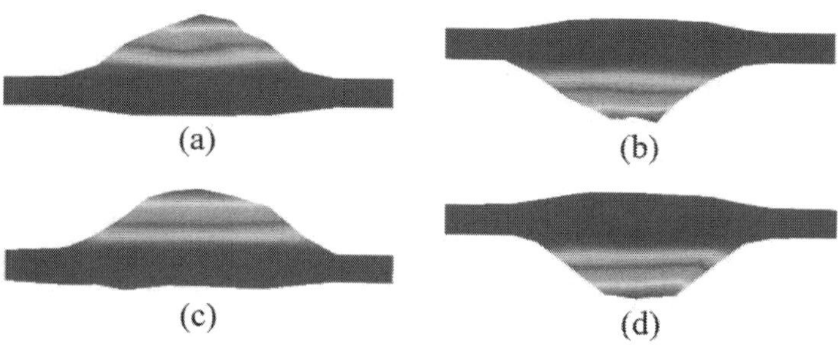

Figure 5. Superelastic cymbal transducer (a) asymmetric (23495 hz) and (b) symmetric (32698 hz) modes, and shape-memory transducer (c) asymmetric (23525 hz) and (d) symmetric (25650 hz) modes from ema. for (a) and (c) the lower end cap is concave, for (b) and (d) the upper end cap is convex. blue to red indicates regions of low to high displacement.

In general, DSC is a reliable and precise technique, and the sample tests in this study were highly repeatable. It is therefore unlikely that there are any systematic DSC errors associated with the data recorded for the superelastic alloy, and this is evidenced by the fact the DSC measured transformation temperatures correlate well with the recorded phase transformation in the impedance analysis of the shape-memory transducer. It has been reported previously that DSC analysis of alloys fabricated for superelastic capability can produce inconclusive results [32], but there is no previous experimental evidence or verification reported in the literature. The data shown in Figs. 3 and 4shows that caution must be exercised when characterizing the material behavior of superelastic nitinol which has been cold-worked, and that the consequent inconclusive interpretation of transformation temperatures has serious implications for the design of tunable transducers based on superelastic nitinol.

SECTION IV

MODAL BEHAVIOR

As with most cymbal transducers, the slight asymmetries associated with transducer fabrication, including small geometrical differences between the end caps and any unevenness in the bond layer, result in the excitation of both the symmetric and the asymmetric form of the axial mode. This is commonly observed as a double-peak in the impedance—frequency spectrum [7], [33], as seen in Fig. 4. To verify the presence of these modes, experimental modal analysis (EMA) was conducted using a 3-D LDV for both transducers in the low-temperature R-phase state. The measured axial modes of each transducer are shown in Fig. 5.

The modes are confirmed as the asymmetric and symmetric axial modes. The asymmetric mode is characterized by in-phase axial motion of the two end caps, whereas the symmetric mode is characterized by out-of-phase axial motion of the end caps. However, for both modes, one of the end caps is dominant in the response. This is significant for incorporation of this transducer into a transducer, because for each resonant frequency only one of the end caps provides the axial resonant displacement amplitude that could be exploited. It is therefore important to understand the modal behavior of such devices, but also the vibration response of each mode, in particular the achievable displacement amplitudes.

SECTION V

VIBRATION RESONANCE RESPONSECHARACTERIZATION

The vibration displacement response in both the low temperature R-phase state and the high-temperature austenite phase in a frequency range around resonance in each mode was measured for both transducers, with the resonant frequencies initially as identified from the impedance analysis (Fig. 4) and EMA (Fig. 5). The two transducers were excited with a burst-sine excitation signal in bidirectional frequency sweeps through each resonance. Transducers tend to heat up as a result of extended durations of excitation during vibration characterization measurements and the consequent changes in resonant frequencies can easily be misinterpreted as nonlinear responses. The temperature of the transducers in this study also had be carefully monitored to prevent an uncontrolled phase transformation. To minimize heating in the R-phase, and to maintain consistency throughout the experiments, burst-sine signals were used with 2 to 4 s durations between bursts for all measurements, with the length of time required being dependent on the voltage level, which ensured that the cymbal transducers were maintained at room temperature. The choice of bidirectional frequency sweeps for the characterizations allowed for nonlinearities in the responses not associated with heating to be identified from any differences between the sweep-up and sweep down responses. Experiments were conducted at five voltage levels, incremented from 2 V to 10 V. The results for the superelastic and shape-memory cymbal transducers are shown in Figs. 6 and 7, respectively.

As can be observed in Figs. 6 and 7, the measured resonant frequencies differ marginally from the resonant frequencies measured by impedance analysis and EMA. This is due, in part, to the fact that the transducers are being driven at higher voltages, meaning there are small changes in how the nitinol responds. This change in behavior has been identified in cymbal transducers fabricated using non-SMA materials such as brass [15], but it is important to clarify that nitinol is very sensitive to both temperature and strain [34], and so the higher strains excited in these experiments result in a small shift in resonant frequency greater than those resulting from much lower strain electrical impedance measurements.

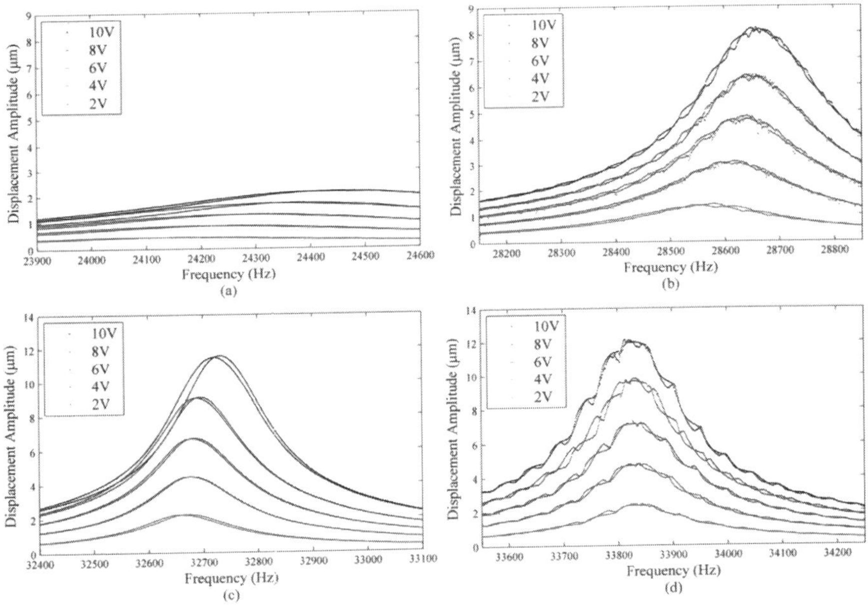

Figure 6. Vibration responses of the asymmetric mode in the (a) r-phase (24°c) and (b) austenite phase (60°c), and symmetric mode in the (c) r-phase (24°c) and (d) austenite phase (60°c) of the superelastic transducer.

A. Limitations of the Cymbal Transducer Configuration

The results show that the phase transformation associated with each mode correlates well with the shift in resonant frequency identified in the impedance analysis. However, for both asymmetric and symmetric axial modes, lower displacement amplitude is exhibited in the R-phase compared with the austenite phase. For a pure phase transformation, one would expect the transition to a stiffer phase, for example from martensite or R-phase to austenite, to result in resonance at a higher frequency accompanied by lower displacement amplitude. However, the results shown in Figs. 6and 7 indicate that, in some circumstances, the transducers exhibit behavior inconsistent with this premise, resulting in a higher displacement amplitude and a lower resonant frequency for the shape-memory nitinol transducer asymmetric mode. Although this appears to be counterintuitive, it is important also to consider that any heating, and therefore softening, of the epoxy layer could be a major influencing factor in the measured responses. The epoxy layer is, in fact, a weakness in the classical design of cymbal transducers for higher power and higher displacement applications and different assembly configurations have been researched to overcome this limitation [15], [16].

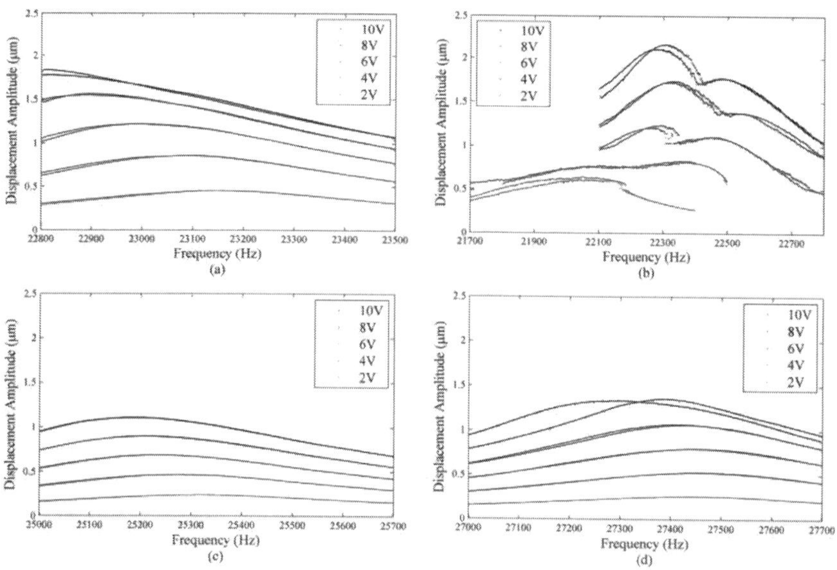

Figure 7. Vibration responses of the asymmetric mode in the (a) r-phase (22°c) and (b) austenite phase (70°c), and symmetric mode in the (c) r-phase (24°c), and (d) austenite phase (80°c) of the shape-memory transducer.

There is further evidence that the high-temperature environment is influencing the epoxy bond layer behavior. Modulations in the frequency response are noticeable for both modes in the austenite phase in the results shown in Fig. 6. These modulations are not unique to SMA and have been similarly identified for steel-titanium cymbal transducers. It is known that raised temperatures lower the flexural and storage moduli and also the flexural strength of epoxy resin adhesives [35] and the measured responses are consistent with loss of structural integrity of the epoxy. In Fig. 7, for both modes in the higher temperature phase, the modulated distortions of the displacement response are not observed but the responses are at much lower amplitudes. This is evidence that the bond layer integrity is influenced by a combination of temperature and strain. It is worth noting that the higher temperature phase measurements are conducted close to the operational limit of the epoxy resin, which is quoted at 90°C.

B. Tunable Frequency and Nonlinear Behavior

The resonant frequency changes through a phase transformation exhibited for both the asymmetric and symmetric modes for each transducer were consistent with the electrical impedance measurements. It has been demonstrated that the superelastic transducer can be tuned via a change in temperature. However, it has not been possible to achieve a purely stress-induced martensitic transformation in the superelastic region to take advantage of the superelastic effect. This means that the stresses in the end caps of the superelastic transducer are not sufficiently high in this study. This can be investigated in further studies by exciting at much higher input voltages, but this would require a different cymbal design to avoid break-down of the epoxy layer. The shape-memory nitinol cymbal transducer was also successfully tuned, but it is evident that the asymmetric mode has experienced a split in the resonance peak in the austenite condition, as seen in Fig. 7(b), further evidence that the epoxy layer is compromised.

Nonlinear vibration responses have also been identified in the measurements of both cymbal transducers. Softening nonlinearity, in which the resonant frequency decreases as the voltage (and hence displacement amplitude) increases, is detected in the shape-memory transducer (Fig. 7), whereas the superelastic transducer exhibits global hardening (Fig. 6), in which the resonant frequency increases with voltage level. For the cymbals tested in this study, these frequency shift were very small, around 1 %, but the shifts would be expected to be larger and more influential on transducer performance in high-power devices.

SECTION VI

IMPLICATIONS FOR TUNABLE TRANSDUCERDESIGN

Two cymbal transducers have been designed and assembled which can change resonant frequency under a modest temperature increase. However, there are important design considerations which are apparent. There is a restriction of the DSC technique when characterizing superelastic material, and there are implications for transducer design using this type of alloy. For the design of devices to be operated in the austenite phase around room temperature, a superelastic alloy would ideally be selected because of the low temperature. However, characterization of cold worked superelastic nitinol is shown to be unreliable using DSC alone. If the transformation temperatures cannot be identified accurately at the design stage, then adoption of superelastic nitinol in tunable devices can only rely on experimental characterization of prototypes, and therefore there must be considerable leeway in the tuned frequency requirements. For applications permitting the device to be driven at higher temperatures, shape-memory nitinol is an appropriate choice.

The relationship between stress and transformation temperature of nitinol is well known [17], [20]; however, this relationship is not understood when the end caps are operating under cyclic stress conditions at high displacement amplitude. If, for example, the nitinol is heated to the austenite phase and held at a single temperature, a sufficiently high stress could cause the microstructure to reorient to R-phase or martensite. This has implications for the use of nitinol in high-power ultrasonic cymbal transducer end caps. As the stress increases, the temperature must increase to ensure a stable transformation into the higher temperature phase. Therefore, even at the high stress levels associated with high displacement amplitude, it is possible that the nitinol has not completely phase transformed and an increase in temperature is additionally required, or even higher stress.

Exploration of alternative SMAs would also be beneficial in the development of tunable cymbal transducers using active materials, which may prove to exhibit more favorable properties than nitinol. An example is Zn45Au30Cu25 [36] which has been recently developed and shown to exhibit improved reversibility compared with nitinol, as well as lower thermal hysteresis and enhanced thermal cyclic repeatability.

SECTION VII

CONCLUSIONS

Tunable frequency behavior of cymbal transducers incorporating nitinol end caps has been identified and characterized using both thermal and dynamic measurement techniques. The temperature dependency of the elastic properties of nitinol can be exploited to tune the resonant frequency of a cymbal flextensional transducer. For a cymbal transducer with shape-memory end caps, it was shown that DSC of the nitinol alloy provided accurate estimations of the transformation temperatures that could be exploited to design a transducer at known tuned frequencies. However, the transformation temperatures at which a change in frequency was measured for the transducer with superelastic nitinol end caps did not correlate well with the temperatures estimated from DSC analysis. The problems identified with characterization of superelastic nitinol, particularly where it has been cold-worked, impose limitations for the design of cymbal transducers incorporating superelastic nitinol.

There is evidence of nonlinear vibration responses associated with the asymmetric and symmetric axial modes in the austenite phase and R-phase for both transducers. It is also clear that the standard cymbal transducer design is not suitable for high-power ultrasonics applications because of loss of integrity of the epoxy layer. High stress changes the phase transformation temperatures of nitinol alloys but the effects of the cyclic stresses excited in ultrasonic transducers are unknown. Nevertheless, there is potential for the design of tunable devices using active materials such as nitinol and the opportunities for tuning the operating frequency of a device through small changes in temperature are considerable.

FOOTNOTES

This work has been funded by the Engineering and Physical Sciences Research Council (EPSRC) grant EP/P505534/1. The authors also acknowledge the EPSRC Engineering Instrument Loan Pool, which provided the differential scanning calorimeter

REFERENCES

1. J. Zhang, W. J. Hughes, R. J. Meyer, K. Uchino and R. E. Newnham "Cymbal array: A broadband sound projector" *Ultrasonics, vol. 37, no. 8, pp. 523-529, 2000*

2. J. Zhang, A.-C. Hladky-Hennion, W. J. Hughes and R. E. Newnham "Modeling and underwater characterization of cymbal transducers and arrays" *IEEE Trans. Ultrason. Ferroelectr. Freq. Control, vol. 48, no. 2, pp. 560-568, 2001*

3. J.F. Tressler, R.E. Newnham and W.J. Hughes "Capped ceramic underwater sound projector: The ?cymbal? transducer" *J. Acoust. Soc. Am., vol. 105, no. 2, pp. 591-600, 1999*

4. J. F. Tressler, W. Cao, K. Uchino and R. E. Newnham "Finite element analysis of the cymbal-type flextensional transducer" *IEEE Trans. Ultrason. Ferroelectr. Freq. Control, vol. 45, no. 5, pp. 1363-1369, 1998*

5. A.Dogan and E. Uzgur "Size and material effects on cymbal transducer for actuator applications" *Ferroelectrics, vol. 331, no. 1, pp. 53-63, 2006*

6. R. E. Newnham, J. Zhang and R. Meyer "Cymbal transducers: A review" *Proc. IEEE Int. Symp. Applications of Ferroelectrics., pp. 29-32,*

7. R. J. Meyer, W. J. Hughes, T. C. Montgomery, D. C. Markley and R. E. Newnham "Design of and fabrication improvements to the cymbal transducer aided by finite element analysis" *J. Electroceram., vol. 8, no. 2, pp. 163-174, 2002*

8. R. E. Newnham, J. Zhang, S. Alkoy, R. Meyer, W. J. Hughes, A. C. Hladky-Hennion, J. Cochran and D. Markley "Cymbal and BB

underwater transducers and arrays" *Mater. Res. Innovat., vol. 6, no. 3, pp. 89-91, 2002*

9. J. Zhang, W. J. Hughes, A. C. Hladky-Hennion and R. E. Newnham

10. "Concave cymbal transducers" *Mater. Res. Innovat., vol. 2, no. 5, pp. 252-255, 1999*

11. R. E. Newnham, S. Alkoy, A. C. Hladky, W. J. Hughes, D. C. Markley, R. J. Meyer and J. Zhang "Underwater flat-panel transducer arrays"*OCEANS 2001 MTS/IEEE Conf. and Exhibition, vol. 3, pp. 1529-1535,*

12. N. B. Smith, S. Lee, E. Maione, R. B. Roy, S. McElligott and K. K. Shung "Ultrasound-mediated transdermal transport of insulin in vitro through human skin using novel transducer designs" *Ultrasound Med. Biol., vol. 29, no. 2, pp. 311-317, 2003*

13. E. J. Park, J. Werner and N.B. Smith "Ultrasound mediated transdermal insulin delivery in pigs using a lightweight transducer" *Pharm. Res., vol. 24, no. 7, pp. 1396-1401, 2007*

14. K. H. Lam, X. X. Wang and H. L. W. Chan "Lead-free piezoceramic cymbal actuator" *Sens. Actuators A, vol. 125, no. 2, pp. 393-397, 2006*

15. C. Mo, D. Arnold, W. C. Kinsel and W. W. Clark "Modeling and experimental validation of unimorph piezoelectric cymbal design in energy harvesting" *J. Intell. Mater. Syst. Struct., vol. 24, no. 7, pp. 828-836, 2012*

16. F. Bejarano, A. Feeney and M. Lucas "A cymbal transducer for power ultrasonics applications" *Sens. Actuators A, vol. 210, pp. 182-189, 2014*

17. S. Lin "An improved cymbal transducer with combined piezoelectric ceramic ring and metal ring" *Sens. Actuators A, vol. 163, no. 1, pp. 266-276, 2010*

18. S. A. Thompson "An overview of nickel-titanium alloys used in dentistry" *Int. Endod. J., vol. 33, no. 4, pp. 297-310, 2000*

19. T. A. Hall "Joint, a laminate, and a method of preparing a nickel-titanium alloy member surface for bonding to another layer of metal" *1993*

20. J. A. Shaw, C. B. Churchill and M. A. Iadicola "Tips and tricks for characterizing shape memory alloy wire: Part 1-Differential scanning calorimetry and basic phenomena" *Exp. Tech., vol. 32, no. 5, pp. 55-62, 2008*

21. P. K. Kumar and D. C. Lagoudas Shape Memory Alloys-Modeling and Engineering Applications *pp. 1-51, 2008, Springer*

22. X. Huang and Y. Liu "Effect of annealing on the transformation behavior and superelasticity of NiTi shape memory alloy" *Scr. Mater., vol. 45, no. 2, pp. 153-160, 2001*

23. P. P. Poncet "Nitinol medical device design considerations" *Strain, vol. 2, no. 4, pp. 1-12, 2000*

24. V. V. Rubanik, V. V. Rubanik and V. V. Klubovich "The influence of ultrasound on shape memory behavior" *Mater. Sci. Eng. A, vol. 481, pp. 620-622, 2008*

25. S. Belyaev, A. Volkov and N. Resnina "Alternate stresses and temperature variation as factors of influence of ultrasonic vibration on mechanical and functional properties of shape memory alloys" *Ultrasonics, vol. 54, no. 1, pp. 84-89, 2014*

26. R. J. Meyer and R. E. Newnham "Flextensional transducers with shape memory caps for tunable devices" *J. Intell. Mater. Syst. Struct., vol. 11, no. 3, pp. 199-205, 2000*

27. A.B. Flatau, M. J. Dapino and F. T. Calkins "High bandwidth tunability in a smart vibration absorber"

28. *J. Intell. Mater. Syst. Struct., vol. 11, no. 12, pp. 923-929, 2000*

29. M. C. Sharp "Ultrasonic dental scaler selectively tunable either manually or automatically" *2001*

30. P. Fenton, F. Harrington and P. Westhaver "Disposable ultrasonic soft tissue cutting and coagulation systems" *2003*

31. K. A. Williams, G. T.-C. Chiu and R. J. Bernhard "Passive-adaptive vibration absorbers using shape memory alloys" *Proc. SPIE, vol. 3668, pp. 630-641,*

32. Standard Specification for Wrought Nickel-Titanium Shape Memory Alloys for Medical Devices and Surgical Implants *2005*

33. P. Ochoa, J. L. Pons, M. Villegas and J. F. Fernandez "Effect of bonding layer on the electromechanical response of the cymbal metal-ceramic piezocomposite" *J. Eur. Ceram. Soc., vol. 27, no. 2, pp. 1143-1149, 2007*

34. N. A. Obaisi "Determination of the transformation temperature ranges of orthodontic nickel-titanium archwires" *2013*

35. P. Ochoa, J. L. Pons, M. Villegas and J. F. Fernandez "Advantages and limitations of cymbals for sensor and actuator applications" *Sens. Actuators A, vol. 132, no. 1, pp. 63-69, 2006*

36. T. DuerigASM Handbook, Volume 23: Materials for Medical Devices*pp. 237-250, 2012, ASM International*

37. Harismendy, R. Miner, A. Valea, R. Llano-Ponte, F. Mujika and I. Mondragon"Strain rate and temperature effects on the mechanical behaviour of epoxy mixtures with different crosslink densities"*Polymer (Guildf.), vol. 38, no. 22, pp. 5573-5577, 1997*

38. Y. Song, X. Chen, V. Dabade, T. W. Shield and R. D. James "Enhanced reversibility and unusual microstructure of a phase-transforming material" *Nature, vol. 502, no. 7469, pp. 85-88, 2013*

INDEX